W9-BKX-548

GEOMETRY
Fundamental Concepts and Applications

ALAN BASS

PEARSON
Addison
Wesley

Boston   San Francisco   New York
London   Toronto   Sydney   Tokyo   Singapore   Madrid
Mexico City   Munich   Paris   Cape Town   Hong Kong   Montreal

Senior Acquisitions Editor: Jennifer Crum
Editorial Assistant: Antonio Arvelo
Senior Managing Editor: Karen Wernholm
Senior Production Supervisor: Tracy Patruno
Senior Designer: Barbara T. Atkinson
Photo Researcher: Beth Anderson
Marketing Manager: Jay Jenkins
Senior Technology Specialist: Joe Vetere
Rights and Permissions Advisor: Dana Weightman
Manufacturing Manager: Evelyn Beaton
Cover and Text Design: Sandy Silva
Production Coordination: Windfall Software
Composition and Illustrations: LM Graphics

Cover image: © Vincent Burkhead Studio

Photo credits: p. 1, Corbis Royalty Free, MedioImages/Getty Royalty Free; p. 3, Vincent
Burkhead Studios; p. 19, PhotoDisc; p. 37, PhotoDisc; p. 40, courtesy of author; p. 55,
illustration from Jules Verne's *Mysterious Island* (1875); p. 70, courtesy of author; p. 71,
PhotoDisc; p. 77, PhotoDisc Red, Digital Vision, p. 78, Pearson Learning Group, Digital
Vision; p. 83, Michael Drechsler Jewelry Ltd., Steven L. Martin (University of California,
Berkeley); p. 86, PhotoDisc; p. 89, Digital Vision, PhotoDisc Blue, Corbis; p. 93, USGS
Terraserver; p. 105, NationalAtlas.gov; p. 109, EMG Education Management Group; p. 112,
The Bridgeman Art Library International; p. 117, courtesy of Beth Anderson; p. 119, courtesy
of Vincent Burkhead Studios, PhotoDisc ; p.125, PhotoDisc; p. 129, PhotoDisc; p. 130, Corbis
Royalty Free; p. 138, PhotoDisc; p. 141, NASA; p. 151, Corbis Royalty Free; p. 151, courtesy
of author

Cataloging-in-Publication Data on file in the Library of Congress.

ISBN-10: 0-321-47331-0
ISBN-13: 978-0-321-47331-8

1 2 3 4 5 6 7 8 9 10—BB—11 10 09 08 07

Contents

Preface

Welcome to ***Geometry: Fundamental Concepts and Applications.*** This book was created to make geometry accessible and exciting for both students and instructors. It can be used as a supplement to basic math and algebra books that do not contain enough geometry or as the main textbook for a short 1–2 credit hour geometry course.

This book has three main goals. First and foremost, we want to make geometry accessible and interesting for students who may have had trouble with it in the past. We have designed many of the features in the book with this in mind. For the writing style, we have taken a relaxed tone and focused on an intuitive explanation of concepts, techniques, and even formulas. To help establish historical relevance, we begin the book with a brief "Section H" that explores the history of geometry. In addition, we periodically include Historic Note boxes throughout the text. To help establish real-world context, we have made sure that the many applications included are genuinely practical.

As our second goal, we want to provide a geometry text that can be used to review fundamental concepts, but is at a level appropriate for college students. This means presenting examples and homework problems that require a deeper understanding of concepts and an ability to use geometry along with other mathematical skills. We want students to recognize the connection between geometry and other branches of mathematics, particularly algebra. This text incorporates a variety of basic algebra skills while keeping the focus on the students' understanding of geometry.

Finally, we want a text that provides a wide range of pedagogical options for instructors. The text is designed so that instructors can do a cursory review or, if they choose, go as far as incorporating constructions and formal proofs. These features are integrated into the text in a way that makes their inclusion completely optional.

Below is a list of the features that are included throughout the text. We hope that these features, along with the relaxed but mathematically precise language, help students better understand and enjoy the course.

Historical Note Boxes: These boxes, highlighted with ✻, are used to show students the rich history of geometry. Instead of just including facts about who did what when, we have also included information that will help the students grasp of concepts by exploring such things as different civilizations' approaches to the study of geometry and how the first close approximation of π came about.

Point of Interest Boxes: Often the most valuable lesson can be learned through a helpful tip or quick note of interest. Point of Interest Boxes, indicated with an exclamation point (**!**), do just this. They provide the student with insight by flagging material that may be important in a future section or by making a quick point about a real-world scenario that makes the topic more relevant.

Good Question Boxes: We have modeled the student-teacher interaction in the classroom by anticipating, asking, and answering questions that a thinking student might have as topics are presented. This also models inquisitive thinking and general curiosity. These boxes are highlighted with a question mark (**?**).

Arithmetic and Geometry/Algebra and Geometry Examples and Exercises: The Arithmetic/Algebra and Geometry Examples and Exercises bridge concepts that are common to arithmetic and geometry or algebra and geometry. They help to bring a visual interpretation to the arithmetic and show a connection between geometry and algebra-two areas of mathematics that are all too often taught in isolation from each other.

Vocabulary Checklist: Because vocabulary is such an integral part of geometry, a Vocabulary Checklist is provided at the beginning of each exercise set so students can assess their understanding of the terms introduced in the section. The Vocabulary Checklist is followed by a set of vocabulary exercises.

Construction Examples and Exercises: This optional material covers common construction techniques. Some use only a straight-edge and compass; others incorporate the use of protractor and ruler. The material is boxed to separate it from the main text, and the exercises appear at the end of homework in their own section label "Construction."

Geometry Projects: Each section concludes with an optional Geometry Project. These cover a wide variety of techniques, from exploring how theorems are proved to finding the height of a building. Material on proofs in geometry has been included as three of the section projects.

For Instructors: An On-line Answer Book is available to you through our Instructor Resource Center Web site. If you need assistance in accessing this site, please contact your local Addison-Wesley representative.

Acknowledgments

I have been more fortunate than I could have possibly hoped or deserved in my professional relationships on this project. The Math Department at San Diego Mesa College was patient and supportive through the development of this material. I am indebted to Addison-Wesley to an extent that cannot be expressed concisely, and everyone I have worked with on their team has been amazing. In particular, thanks to Jennifer Crum, Tracy Patruno, and Antonio Arvelo for their hard work, support, and inspiration. Paul Anagnostopoulos at Windfall Software is just wonderful and, together with Laurel Muller of LM Graphics, they are responsible for so much of what is good being printed and any bad being gone. Thanks to Vincent "Mac" Burkhead for letting us use and abuse his art (which I love). To my wife Holly: lifetime soul mate, perfect match. She edited, did the homework three times, edited some more, and taught me how to do basic geometry proofs. What can I say? This book is dedicated to her.

And to the students and instructors using this book: We wish you all the best. Geometry is such a beautiful subject. Enjoy it!

GEOMETRY

H

A Brief History of Geometry

As early as 25,000 B.C. we find evidence of basic geometric designs being drawn. However, the earliest recorded explorations of geometry come from the Egyptians and the Babylonians around 3000 B.C. Most of the principles used in early geometry were the result of intuitive observation. As an illustration, look at the photos in the margin. The first is of six Egyptian pyramids and the second is an Egyptian courtyard. Notice that you can tell intuitively which angles and lengths would have the same measure. This is the case with many applications in geometry.

The Egyptians used geometry for a number of practical reasons; mainly architecture and dividing land. Each time the Nile River flooded, land would have to be resurveyed and divided into appropriate sizes. As for astronomy, this is an area where the Egyptians' intuitive approach came up short. Making measurements in astronomy requires complicated, precise mathematical calculations. What we see from Egyptian astronomy is mostly naming constellations.

Still, some of the ideas developed by the Egyptians and Babylonians were surprisingly sophisticated, and a modern mathematician might be hard pressed to prove them without the use of advanced mathematics like calculus. The Egyptians had a correct formula for volumes used in the construction of the pyramids and the Babylonians had a trigonometry table (we will explore trigonometry in the last part of this book).

Despite all the practical accomplishments of these ancient civilizations, the geometry we will study in this book is, for the most part, what the Greeks discovered between about 500 B.C. and 600 A.D. The Greeks viewed geometry as the ultimate in perfect reasoning and saw geometry as having a strong connection with philosophical truth. In fact, there is a legend that the door to the philosopher Plato's school bore the inscription "Let none enter here who are ignorant of Geometry." The Greeks placed a heavy emphasis on proving relationships through deduction, taking geometry beyond the trial-and-error–driven empirical approach taken by the Egyptians and Babylonians. In fact, the Greeks emphasized abstract thinking to an extreme; Plato believed that students in geometry should use nothing but a compass and straight edge (marked rulers and protractors were a workman's tools and not worthy of a scholar). While this may seem overboard, Plato recognized that future scientists would need a deep understanding of abstract math to solve problems. He was right; the Greeks' rigorous approach to mathematics very quickly led to the kind of math it takes to find the diameter of the sun and the distances between planets and to work with complex mechanical engineering.

Arguably the most influential historical figure in all of geometry is Pythagoras. Pythagoras spent his life studying math, geometry, music, and philosophy. He discovered and proved the vast majority of what you will learn about in this book. And while both the Egyptians and the Babylonians had intuitively discovered the Pythagorean

Theorem ($a^2 + b^2 = c^2$ in a right triangle), it was given his name because he was the first to actually prove it.

Other famous Greek mathematicians and geometers included Euclid and Archimedes. Euclid was responsible for writing a book that presented the Greeks' ideas on geometry in their ideal abstract form. The book (called *The Elements*) was one of the first elementary textbooks and included definitions and basic geometric principles derived from five basic assumptions (called *axioms*):

1. It takes two points to determine a straight line.
2. Any line segment can be extended into a straight line.
3. A circle can be drawn with any given center and radius.
4. All right angles have the same measure.
5. Given a line, for every point not on that line there is exactly one line that goes through that point that is parallel to the original line.

The Greeks believed these axioms were so self-evident that they needed no proof.

Archimedes is considered by many to be the greatest of the Greek mathematicians and is often named as one of the three greatest mathematicians of all time. He further developed abstract concepts in geometry, but was also very practical. He invented the pulley and war machines to fight the Romans.

During the period of time after the Roman Empire succeeded the Greek states, there was no notable advance in geometry. And from around 600 A.D. into the 1600s, geometry was largely neglected in favor of algebra. This may be because, thanks to the Greeks' extensive discoveries, geometry had reached a limit in its practical application, while algebra showed great potential. The next major development in geometry came in the 17th century when the philosopher and mathematician Rene Descartes developed a way to analyze algebraic equations using geometry (a subject called *analytic geometry*). Every student in beginning algebra will recognize the example in the margin. The rectangular coordinate system provides a connection between geometric shapes and algebraic equations. This connection between two completely different areas of math was an essential component of higher-level math (particularly calculus) and quantitative physics.

Into the 18th century and beyond, geometry has been advanced both for practical applications and as an abstract field.

The algebraic equation $2x - 3y = 6$ can be represented by a straight line graphed on a coordinate system.

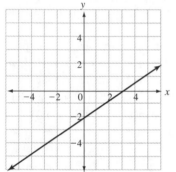

This kind of connection between geometry and algebra is extremely important for higher-level math and is studied in great detail in algebra courses.

HISTORY EXERCISES

1. Discuss the difference in the approach to geometry taken by early civilizations versus the Greeks' approach.

2. According to legend, what inscription did Plato have above the door to his school?

3. What three Greeks are listed in the history as influential figures in the development of geometry?

4. What five axioms did the Greeks base their geometry on and why did they not prove them?

5. What is meant by the term *analytic geometry*?

Basic Geometry Concepts

■ An Overview of This Supplement

Geometry is the branch of mathematics that explores the properties, measurement, and relationships of points, lines, angles, surfaces, and solids. The practical applications that come out of this branch of mathematics range from finding how much concrete will be needed to fill in a foundation to finding the distance between stars. At the same time, geometry is the apex of abstract reasoning and logic; every result can be proven.

Learning basic geometry involves definitions, notations, theorems (results), formulas, constructions, proofs, and using arithmetic to find measurements. This supplement will provide a solid foundation in basic geometry. The constructions and proofs are an optional component and have been separated in the text and homework from the main body of material.

Algebra is the branch of mathematics in which symbols (usually letters of the alphabet) are used to represent numbers that are unknown or changing, and relationships are explored that hold for all numbers. In short, algebra is involved anywhere that variables are used. Popular techniques in algebra include simplifying expressions and solving equations.

Much of this supplement was written with the intention of mixing geometry and algebra. This is not hard to do; these subjects provide a good mix for both abstract problems and practical applications. We will explore concepts and principles from basic geometry—angles, polygons, perimeter, area, volume, surface area, and so on—and whenever possible we will be introducing variables into the examples and exercises in order to reinforce techniques from algebra and explore how they are used with geometry.

We will try to nourish an appreciation for both the logic-driven, abstract parts of geometry and the hands-on, real-world applications it solves. This will give you an idea of how geometry is used to solve some advanced problems in different vocational fields. In this first section, we will explore some of the fundamental elements of geometry. We begin with the concepts of points, lines, and planes.

■ Points, Lines, and Planes

All figures in geometry are made up of points, lines, and planes. These three concepts are so fundamental to geometry that there are no other words to define them. Their existence is accepted and understood intuitively. While we cannot give an official definition, we can discuss some of their features.

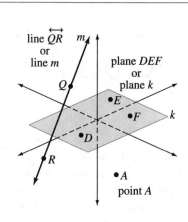

line $\overleftrightarrow{QR}$
or
line m

plane DEF
or
plane k

Q

E

F

k

D

R

A
point A

Features of a point:

- Points have no dimension (zero-dimensional). This means a point has no length, width, or height.
- A point is identifiable only by its location.
- Points are represented by a dot.
- Notation: Points are usually named with a capital letter.

Features of a line:

- Lines are one-dimensional. This means a line has infinite length, but no width or height.
- A line contains an infinite number of points.
- Notation: $\overleftrightarrow{QR}$ represents the line that goes through the points Q and R. Lines can also be named with a lowercase letter.

Features of a plane:

- Planes are two-dimensional. This means a plane has infinite length and width, but no height.
- Planes contain an infinite number of points and an infinite number of lines.
- Notation: Plane DEF represents the plane that contains these three points. Planes can also be named using a lowercase letter.

Other ideas involving points, lines, and planes:

- Any two distinct points determine one (and only one) straight line.
- Two distinct lines that intersect will intersect at a single point.
- Two lines are **parallel** if they exist in the same plane, but do not intersect. Lines that do not exist in the same plane are called **skew lines.**
- The intersection of two distinct planes creates a line.

Intersecting lines   Parallel lines   Skew lines   Intersecting planes

Understanding the concepts of points, lines, and planes leads to the creation of other objects:

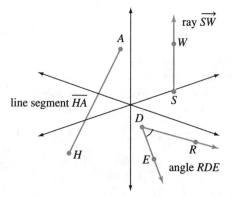

ray $\overrightarrow{SW}$

A

W

line segment $\overline{HA}$

S

D

H

E

R

angle RDE

- A **ray** is the set of all points on a line that start at a point and extend infinitely in one direction. Notation: $\overrightarrow{SW}$ represents the ray that starts at point S and extends in the direction of point W.

- A **line segment** is the set of all points that lie between two distinct points on a line, including the points themselves. Notation: The notation $\overline{HA}$ refers to the line segment itself while the notation HA refers to the length of line segment $\overline{HA}$.

- An **angle** is formed when two rays share an endpoint. We will discuss angles in more detail in Section 2: More about Angles. Notation: $\angle RDE$.

EXAMPLE 1-1   Drawing Basic Geometric Objects

In the figure below, draw the following: $\overrightarrow{DE}$, $\overline{AE}$, $\angle FBH$, $\overleftrightarrow{CG}$

◆ **SOLUTION**

After drawing all the objects listed, the figure will look like the one below. Take a minute to note all the objects created.

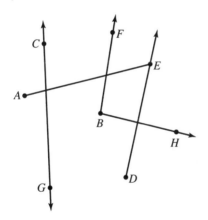

EXAMPLE 1-2   Drawing Geometric Objects

Draw one figure that contains the following objects: $\overrightarrow{GT}$, $\overline{TH}$, D, $\overleftrightarrow{GY}$, $\angle RSW$

◆ **SOLUTION**

Here is one of several possible scenarios for the figure.

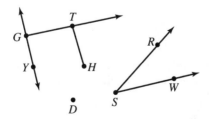

Notice that the points G and T have to be drawn as part of two of the objects asked for, and that point D is not necessarily involved in any other object.   ◆

Here is an example that requires a good understanding of the distinction between a line and a line segment.

How Many Lines? How Many Segments?

How many lines are determined by three distinct points? How many line segments are determined by three distinct points?

◆ **SOLUTION**

For the lines, we need to consider two possible cases.

Case 1: The three points are **collinear** (i.e., they lie on the same line). In this case, there is only one line that can be drawn from the points.

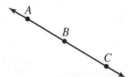

Notice that $\overleftrightarrow{AB}$, $\overleftrightarrow{AC}$, and $\overleftrightarrow{BC}$ all result in the same line.

Case 2: The three points are **noncollinear** (i.e., they do not lie on the same line). In this case, a unique line is determined by every two points.

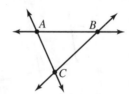

So there are three lines determined by the points.

To see how line segments could be created by three points, notice that whether the three points are collinear or not, the line segments are distinct. That is, in the figure below, the line segments $\overline{AB}$, $\overline{AC}$, and $\overline{BC}$ are all distinct.

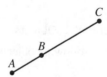

This will be true whether or not the points are collinear. So there are always three line segments determined by three distinct points.   ◆

■ **Between Points**

With regard to points, the term *between* can only be applied to collinear points. So, in the figure in the margin you could say that point P is between points F and Q but P is not between K and S.

Exploring between Points

Refer to the figure in the margin and answer these questions.

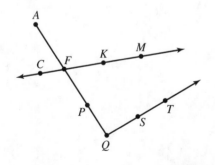

a. Point *S* is between which pair(s) of points?

b. Point *K* is between which pair(s) of points?

c. How many pairs of points is point *F* between?

d. How many pairs of points is point *Q* between?

◆ SOLUTION

a. The only pair of points that *S* is between is *Q* and *T*.

b. The point *K* is between points *F* and *M*. It is also between the points *C* and *M*.

c. The point *F* is between four different pairs of points: *A* and *P*, *A* and *Q*, *C* and *K*, and *C* and *M*.

d. *Q* does not lie between any pair of points.   ◆

As we have mentioned, the notation for a line segment is $\overline{AB}$, while the notation for the length of $\overline{AB}$ is *AB*. We can use this notation to state the definition of *between* more precisely as follows:

In this figure, $AB + BC = AC$ and *B* is between points *A* and *C*.

In this figure, $AB + BC > AC$ and *B* is not between points *A* and *C*.

> Point *B* is **between** points *A* and *C* if $AB + BC = AC$.

The figures in the margin illustrate that the equation $AB + BC = AC$ will only be true if the points are collinear and the point *B* is, indeed, between *A* and *C*.

EXAMPLE 1-5   What We Know

Suppose that three distinct points *D*, *F*, and *K* are arranged so that $DF = 5$, $FK = 7$, and $DK = 9$. In the context of betweenness, what can be said about these points?

◆ SOLUTION

Note that the lengths cannot be arranged in such a way that the sum of two equals the other:

$DF + FK \neq DK$

$DF + DK \neq FK$

$FK + DK \neq DF$

Since the lengths of the line segments cannot be arranged so that the sum of two is equal to the other, we know that no one of these points can be between the other two. ◆

This property of addition for points on the same line segment can be developed further. We can even suggest a property that involves subtraction. Consider the figure in the margin. Here are just a few examples of the kinds of relationships that exist between the lengths of the line segments involved:

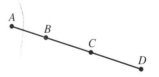

- $AC = AB + BC$

- $BD = BC + CD$

- $AD = AB + BC + CD$

- $BD = AD - AB$

- $AB = AD - BD = AD - (BC + CD) = AD - BC - CD$

EXAMPLE 1-6   The Missing Length

Suppose for the figure below that $AC = 9$, $AB = 4$, and $BD = 11$. Use this information to find CD.

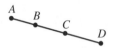

◆ SOLUTION

Since we know $BD = 11$, we could find CD using the subtraction $CD = BD - BC$ if we knew length BC. We can find BC using the other information we were given. That is, $AC = 9$ and $AB = 4$. Now we have

$$AC = AB + BC$$
$$9 = 4 + BC$$
$$BC = 5$$

Now we go back to the equation

$$CD = BD - BC$$
$$CD = 11 - 5$$
$$CD = 6$$
◆

■ Congruent Objects and Bisectors

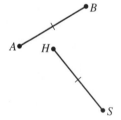

Two objects are called **congruent** if they have the same shape and measure(s). The word *congruent* can be applied to any type of object, so long as it satisfies the definition. The use of tick marks indicates the segments in the margin have the same length. That is, $AB = HS$. Another notation often used is $\overline{AB} \cong \overline{HS}$. These two notations say the same thing: that the line segments are congruent. We will discuss the distinction between the notations further in Section 2 when we explore angles.

To **bisect** an object means to divide it into two congruent parts.

EXAMPLE 1-7   Identifying Congruent Segments and Bisectors

In the figure below, points W, L, and C are collinear, and points K, A, and U are collinear. Indicate which line segments in the figure are congruent and which have been bisected.

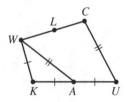

◆ SOLUTION

Three line segments have been marked with a single tick mark, so they are congruent: $\overline{WK} \cong \overline{KA} \cong \overline{AU}$, or $WK = KA = AU$.

Two other line segments have been marked with double tick marks, so they are congruent: $\overline{WA} \cong \overline{CU}$, or $WA = CU$.

Notice the single tick marks and double tick marks indicate different sets of congruent segments.

Also, since points K, A, and U are collinear, and $\overline{KA} \cong \overline{AU}$, line segment $\overline{KU}$ has been bisected by point A.

Notice that it looks like $\overline{WC}$ has been bisected by point L; but we can't be sure since we weren't given marks to indicate $\overline{WL}$ and $\overline{LC}$ are congruent. So $\overline{KU}$ is the only segment we are sure has been bisected.   ◆

In the case of a line segment, the bisector is usually referred to as the **midpoint** of the line segment. At the end of this section we look at how you can construct the midpoint of a line segment without using a measuring ruler. Right now, we just want to understand what the definition implies.

EXAMPLE 1-8   The Midpoint of a Line Segment

If point H is a midpoint for line segment $\overline{SB}$ and $SB = 15$, draw a corresponding figure and label the length of every line segment involved.

◆ **SOLUTION**

Since point H is a midpoint, the total length of $\overline{SB}$ should be divided into two segments of equal length. Since $15 \div 2 = 7.5$, the length of each segment should be 7.5 units.

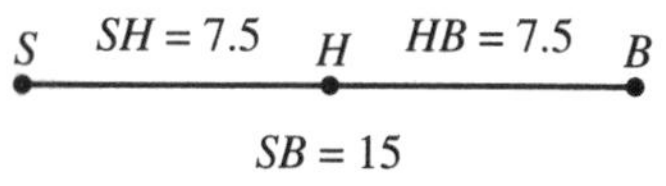

$$S \quad \overset{SH = 7.5}{\bullet\!\!-\!\!-\!\!-\!\!-\!\!-} \overset{H}{\bullet} \overset{HB = 7.5}{-\!\!-\!\!-\!\!-\!\!-\!\!\bullet} \quad B$$
$$SB = 15$$

◆

EXAMPLE 1-9   Midpoints and Algebra

In the figure below, line segments $\overline{ZE}$ and $\overline{EP}$ have algebraic expressions given for their lengths. Find the length of every line segment in the figure given that point E is the midpoint of line segment $\overline{ZP}$.

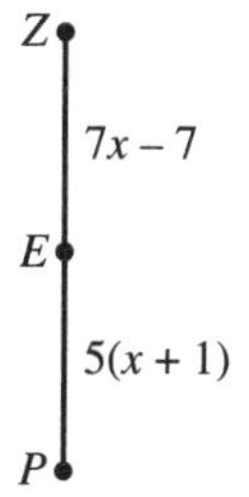

Z
$7x - 7$
E
$5(x + 1)$
P

◆ **SOLUTION**

Since point E is a midpoint for $\overline{ZP}$, it must be true that $ZE = EP$. This leads to the equation

$$7x - 7 = 5(x + 1)$$
$$7x - 7 = 5x + 5$$
$$2x = 12$$
$$x = 6$$

It is a common mistake for students to stop here and say that the solution to the problem is $x = 6$. Notice, however, that we were told to find the length of every line segment in the figure. The variable x is an unknown part of those lengths. The answers are

$$ZE = 7x - 7 = 7(6) - 7 = 35$$
$$EP = 5(x + 1) = 5(6 + 1) = 35$$
$$ZP = 35 + 35 = 70$$

If the lengths ZE and EP had come out different from each other, that would have told us we had made a mistake somewhere.   ◆

■ Solving Word Problems in Geometry

Here is a summary of some of the important strategies that may help in solving geometry word problems:

1. Draw a figure corresponding to the problem whenever possible.
2. Assign a variable or variables and/or expressions to whatever quantity it is you are being asked to find. This should be one of your first steps.
3. Try to summarize the information using as many mathematical symbols as possible.
4. Generate an equation.
5. Solve the equation.
6. Check your answer by making sure it makes sense in the context of the problem.

EXAMPLE 1-10

A piece of wire is to be cut into four sections. The second section is to be twice as long as the first section. The third section is to be 1 inch more than three times the first section. The last section is to be three inches less than six times the first section. If the total length of the wire is 46 inches, what will be the length of each piece?

◆ SOLUTION

The information is much easier to process if we go ahead and draw a wire cut into four parts. This will help us summarize the information and translate it into math.

x = length of the 1st section	2nd section is twice as long as 1st	3rd section is 1 more than 3 times 1st	4th section is 3 less than 6 times 1st
x	$2x$	$1 + 3x$	$6x - 3$

We generate and then solve the following equation:

$$(x) + (2x) + (1 + 3x) + (6x - 3) = 46$$
$$x + 2x + 1 + 3x + 6x - 3 = 46$$
$$12x - 2 = 46$$
$$12x = 48$$
$$x = 4$$

Note that the value of x gives the length of the first section. We can now find the other lengths.

1st section: 4 inches
2nd section: $2(4) = 8$ inches

3rd section: $1 + 3(4) = 13$ inches
4th section: $6(4) - 3 = 21$ inches

Finally, we can do a quick check to make sure these values add to 46 inches.
$4 + 8 + 13 + 21 = 46$ inches total. ✓   ◆

CONSTRUCTION EXAMPLE

Using a Straightedge and Compass to Bisect a Line Segment

An optional part of what you will get out of this supplement is the ability to construct geometric objects both with measuring devices and without them. This requires work with a ruler or straightedge for drawing lines, rays, and segments, a protractor for angles, and a compass for arcs and, later on, circles.

As mentioned in the History section, the Greeks loved geometry! In particular, they enjoyed trying to construct figures accurately without the use of a measuring tool. A good example is the challenge of bisecting a line segment without using a ruler. The following construction is the solution to the challenge.

1. Begin by creating the line segment.

2. Open the compass so that the distance between the point and pencil is at least half the length of the line segment. Draw a wide arc with the point of the compass on one of the line segment's endpoints.

3. Repeat the process for the other endpoint, leaving the compass at the same width.

4. Mark the points where the arcs intersect and use them to create a line segment.

The line segment created by these intersection points bisects the original line segment, so the point where these segments intersect is the midpoint of the original line segment. Notice the use of tick marks to indicate the two parts are congruent.

So why bother? Why not just take out your ruler and see where the line segment is bisected? Remember, the Greeks viewed geometry as the height of reason and logic, not an exercise in measurement. Also, this procedure is practical for use in proofs and in more complicated constructions (as we will see). You will use this procedure to help with constructions in the homework, and we will do similar types of constructions in other sections throughout this supplement.

> **!** The line segment created by the two constructed points is called a *perpendicular bisector*. We will discuss the term *perpendicular* in Section 2.

Section 1 | Exercises

For exercises 1–6, use the terms from the **Vocabulary Checklist** *to fill in the blanks.*

> ✓ **VOCABULARY CHECKLIST:**
> ✓points, lines, and planes   ✓bisector
> ✓line segment   ✓midpoint
> ✓ray   congruent
> angle

1. A(n) _____ is the set of all points on a line that begin at one point and extend infinitely in one direction.

2. Any time two rays share an endpoint it forms a(n) _____.

3. A point that divides a line segment into two equal parts is called a(n) _____ or _____ for the line segment.

4. Three undefined but extremely important terms in geometry are _____, _____, and _____.

5. The set of all points between and including any two points on a line is called a(n) _____.

6. Any two objects that have the exact same shape and size are called _____.

For exercises 7–10, answer TRUE or FALSE.

T | F   7. A line segment could be given a measure, but a ray could not.

T | F   8. Every line segment has a bisector.

T | F   9. Two rays that share an endpoint will form a straight line.

T | F 10. Given any three points, there is exactly one line that can be drawn through all three.

11. In the figure below, draw these objects: $\angle ABG$, $\overrightarrow{GF}$, $\overrightarrow{AC}$, $\overleftrightarrow{DE}$.

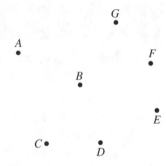

12. Draw one figure that contains all of the following objects: $\angle XWV$, $\overline{RT}$, $\overleftrightarrow{YZ}$, $\overrightarrow{ZX}$, U.

13. Name the following geometric objects using appropriate notation.

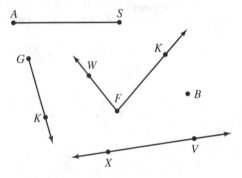

14. The **vertex** of an angle is the point shared by each of its sides. The figure below contains a line, a ray, and a line segment. Sketch the figure on your own paper. Name every angle shown in the figure that has either point A or point S as a vertex.

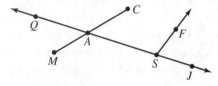

15. Label the points to the figure below so that the following objects have been created:
    $\overline{XY}, \angle XYW, \overrightarrow{TV}, \overline{RS}$.

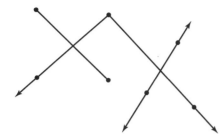

16. Name every line segment in the figure below.

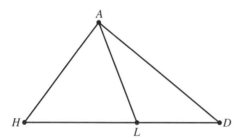

17. Name every line segment in the figure below.

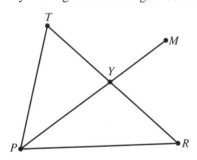

18. Suppose points A, B, and C lie on the same plane. Line segment $\overline{AD}$ also lies on the plane. Draw a figure to show why line segment $\overline{BD}$ must also be in the plane.

19. Is it possible for one point to lie on three different lines? If so, draw a picture of it.

20. Can three distinct points lie on the same line? If so, draw a picture of it.

21. Can two distinct lines go through the same two distinct points? If so, draw a picture.

22. Given line $\overleftrightarrow{XY}$, how many points on the line are exactly 3 inches away from point X? Explain.

23. Given $\overrightarrow{XY}$, how many points on the ray are exactly 3 inches away from point X? Explain.

In the figure below, the line segments form a box. Exercises 24–29 reference this figure.

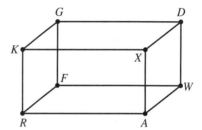

24. Name two line segments that are parallel to line segment $\overline{DX}$.

25. **Coplanar** objects lie on the same plane. Give four points that are coplanar, if possible.

26. **Collinear** points lie on the same line. Give three collinear points, if possible.

27. Name two line segments that are coplanar.

28. Name two line segments that are non-coplanar.

29. How many angles are in the figure?

30. How many lines can be drawn from four distinct points if three of the points are collinear? Draw a figure to illustrate.

31. How many line segments can be drawn from four distinct points if three points are collinear? Draw a figure to illustrate.

32. How many lines can be drawn from four distinct points if no three points are collinear? Draw a figure to explain.

33. How many line segments can be drawn from four distinct points if no three points are collinear? Draw a figure to explain.

Exercises 34–42 refer to the following figure.

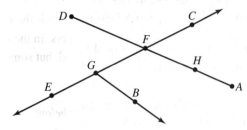

34. Which point(s), if any, are between points *E* and *C*?

35. Which point(s), if any, are between *A* and *D*?

36. Which point(s), if any, are between *G* and *B*?

37. Which pair(s) of points is *G* between?

38. Which pair(s) of points is *H* between?

39. How many pairs of points is *F* between?

40. How many pairs of points is *E* between?

41. Given that point *A* is between *D* and *F*, could the following information be accurate and why?
    $DF = 5, FA = 7, DA = 10$

42. Given that point *G* is between *E* and *C*, could the following information be accurate and why?
    $EC = 13, EG = 9, GC = 4$

43. If point *C* is a midpoint (bisector) for line segment $\overline{UW}$, and $UW = 27.4$, then what is *UC*?

44. If point *T* is a midpoint (bisector) for line segment $\overline{EQ}$, and $EQ = 6\frac{2}{5}$, then what is *ET*?

Exercises 45–52 involve the following figure.

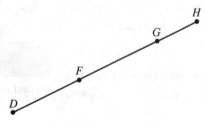

45. Find *GH* if $FG = 10, DF = 6$, and $DH = 19$.

46. Find *DF* if $DH = 40, FG = 23$, and $GH = 7$.

47. Find *DH* if $DG = 7$ and $GH = 2$.

48. Find *FG* if $DF = 4, GH = 3$, and $DH = 15$.

49. Find *GH* if $DG = 15, FH = 13$, and $DF = 6$.

50. Find *DF* if $DG = 27, FH = 23$, and $GH = 7$.

51. What's wrong with the following information?
    $DG = 12, DF = 4, GH = 3, DH = 18$

52. What's wrong with the following information?
    $DH = 20, DF = 6, DG = 16, GH = 6$

53. Given that point *E* is the midpoint of line segment $\overline{LD}$, find *LE*, *ED*, and *LD*.

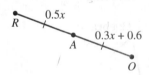

54. Find *RA*, *AO*, and *RO*.

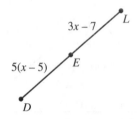

55. Indicate which line segments are congruent.

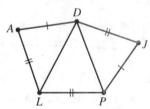

56. Indicate which line segments are congruent.

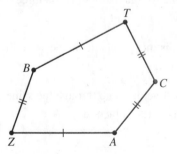

57. **Geometry and Arithmetic.** Given the figure, find the total length of the indicated line segments: $\overline{AC}, \overline{CE}, \overline{AE}$

$$\underset{A}{\bullet} \quad \overset{1.5\ in.}{\underset{B}{\bullet}} \quad \overset{1\frac{3}{4}\ in.}{\underset{C}{\bullet}} \quad \overset{2.1\ in.}{\underset{D}{\bullet}} \quad \overset{\frac{4}{5}\ in.}{\underset{E}{\bullet}}$$

58. **Geometry and Arithmetic.** Sketch a figure corresponding to the following situation and solve the problem:

A line segment is divided into three parts. The first part is $2\frac{2}{3}$ cm. The second part is $\frac{1}{2}$ cm less than three times the first part. The third part is $4\frac{1}{5}$ cm more than the first part. Find the total length of the line segment.

59. **Geometry and Arithmetic.** Sketch a figure corresponding to the following situation and solve the problem:

A line segment is divided into four parts. The first part is 1.5 in. longer than the second part. The second part is 5.7 in. The third part is 0.8 in. shorter than the first part. The fourth is half as long as the other three parts combined. Find the total length of the line segment.

60. **Geometry and Algebra.** A line segment that is 23 feet is to be divided into three pieces. The second piece is to be 4 feet less than twice the first piece, and the third piece is to be twice as long as the second piece. What will be the length of each piece?

61. **Geometry and Algebra.** A line segment that is 43 inches is to be divided into three pieces. The second piece is to be 7 inches less than the first piece, and the third piece is to be twice as long as the second piece. What will be the length of each piece?

Geometry and Graphing

One of the most common uses of lines in math is the **number line**—a line used to represent all real numbers. In the number line below, several points have been graphed, but some information is missing.

Exercises 62–66 reference the number line below.

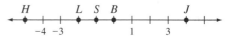

62. What are the coordinates of points H, L, and J?

63. Which point is at 0?

64. What is the length of line segment LJ?

65. Is it true that $\overline{HB} \cong \overline{SJ}$?

66. Which line segment is congruent to $\overline{BJ}$?

67. **Geometry and Graphing.** A and B represent the coordinates of two points on a number line and C represents the coordinate of the midpoint of the points A and B. Find the missing value.

a. $A = 4$, $B = 9$, $C = ?$

b. $A = ?$, $B = 3$, $C = 6$

c. $A = 1\frac{1}{2}$, $B = ?$, $C = 5\frac{3}{4}$

68. A picture 8 inches wide needs to be hung on a segment of a wall that is only $15\frac{1}{3}$ inches wide. If the picture needs to be centered horizontally, how far from the edge of the wall should the picture begin? Draw a divided line segment and label the lengths to illustrate the solution to the problem.

CONSTRUCTION

Some of the homework problems in this supplement require the use of a protractor, a ruler that measures inches and centimeters, and/or a compass. They appear within this section at the end of Exercises.

69. **Ruler.** Construct any ray $\overrightarrow{KJ}$. Construct a line segment $\overline{KL}$ so that the point L is on $\overrightarrow{KJ}$ and $KL = 2\frac{3}{4}$ inches.

70. **Ruler.** Construct a line segment $\overline{AB}$ that is 1.5 inches. Construct a line ℓ so that point A is on ℓ but B is not.

Construct a line segment $\overline{CD}$ on line ℓ so that $\overline{CD}$ is twice the length of $\overline{AB}$ and point A bisects $\overline{CD}$.

71. **Ruler and Compass.** Use a ruler to draw a line segment that is exactly 5 inches. Use the technique outlined in this

section for bisecting a line segment with a compass to bisect the line segment. Now use the ruler to verify that the bisector is in the right position.

72. **Ruler and Compass.** Use a ruler to determine if the line segments below are congruent. Use a compass only to determine if line segment $\overline{TS}$ bisects line segment $\overline{PQ}$ and if line segment $\overline{PQ}$ bisects line segment $\overline{TS}$.

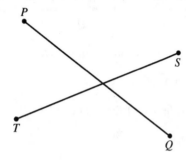

73. **Straightedge and Compass.** Draw any line segment and call it $\overline{AB}$. Now make an exact copy of it using the following steps:

(a) Draw a ray separate from the line segment. Call it $\overrightarrow{HS}$.

(b) Put the point of the compass on point A and open the compass to the width of the segment and place the pencil on point B.

(c) Leaving the compass at the same opening, put the tip on point H and draw an arc that intersects the ray.

(d) Label the point where the arc intersects the ray point L.

(e) Line segments $\overline{AB}$ and $\overline{HS}$ are congruent, which means they have the same length.

Geometry Project 1 | The History of Geometry

At the beginning of this supplement, we very briefly discussed the history of geometry. You will write a report to help expand on that history. There are limitless topics you can pick for geometry. Here are a few examples of things you could write about.

- Pick a civilization or culture and write about their contributions to the body of knowledge in geometry. What new ideas did they develop and what old ideas did they further develop?

- Before the Greeks, geometry was only done for the sake of application. Write about some of the practical applications of geometry that preceded the rigorous academic approach taken by the Greeks.

- Pick a Greek geometer and write about his contributions to early geometry. You can pick a geometer mentioned in the History section or choose another one.

- What is the "Golden Ratio"? When was it first recognized and where? What is its relevance to geometry, nature, science, and art?

More about Angles

Angles are everywhere:

• The Greeks used angles to find the radius of the Earth.

• Roads are graded at angles for proper drainage.

• Airplanes land at angles.

• The photo in the margin shows how a bridge can be supported using angles.

• An entire branch of mathematics (called *trigonometry*) is founded on the study of angles.

In this section we are going to make sure that we understand the fundamentals of how angles are measured and classified. We will also explore certain relationships between angles that are always true and very common in geometry.

Angles can be named several ways. Review the figure of the angle below carefully.

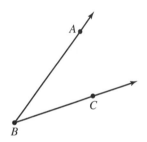

All of the following names are referring to the same angle: ∠ABC, ∠CBA, ∠B.

The point shared by both rays is called the **vertex** of the angle. In this figure, the point *B* is the vertex of the angle.

?

GOOD QUESTION:

Does an angle have to involve two rays?

Answer: In Section 1 we defined an angle as an object created when two rays share an initial point. Practically, however, an angle is created whenever two lines, rays, or line segments share a point. The figure below contains a line, a line segment, a ray, and several angles.

■ Measuring and Classifying Angles

Angles are measured using degrees (using the symbol ° to denote the units). One degree is created by dividing a complete revolution into 360 equal pieces. So one degree looks like this:

There are 360° in a full circle and 180° in a straight line.

Angles can be classified into one of four categories according to their degree measure.

- **Acute angles** measure between 0° and 90°.

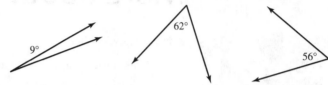

> Notice a "corner" is used to indicate that an angle is exactly 90°. It is a common mistake for students to assume an angle is a right angle when in fact it may not be if this symbol is not used.

- **Right angles** measure exactly 90°.

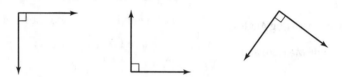

- **Obtuse angles** measure between 90° and 180°.

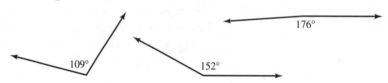

- **Straight angles** measure exactly 180°.

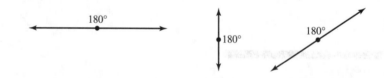

EXAMPLE 2-1   Classifying the Angles in a Quadrilateral

The following figure is called a *quadrilateral* (because it has four sides). We will study quadrilaterals in great detail in Section 4. In the quadrilateral below, name every angle and classify it as acute, right, or obtuse. Assume the angles are drawn to scale.

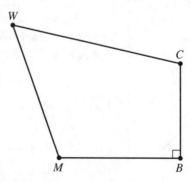

◆ **SOLUTION**

There are four angles in the figure:

- ∠*WMB* is an obtuse angle since it is obviously greater than 90°.

- ∠*MBC* is a right angle since it is marked with a corner.

- ∠*BCW* is an obtuse angle.

- ∠*MWC* is an acute angle since it is obviously less than 90°.

Note that we could have just as easily named the four angles ∠*W*, ∠*M*, ∠*B*, and ∠*C*.

◆

We use a **protractor** to measure angles. Notice that a protractor has a circular scale from 0° to 180°. Notice also that it gives these measurements both from left to right and from right to left. This is necessary because you could be presented with an angle whose initial side points either way.

To use a protractor:

1. Place the center of the straight edge of the protractor on the vertex of the angle.

2. Line the straight edge of the protractor along one side of the angle.

3. Note which scale should be used. Again, the side along the straight edge points to 0°.

4. The second side of the angle will give the angle's measurement.

Of the two angles shown below, in the left angle, it would be most convenient to use the inner scale of a protractor to get a measure of 48°. In the angle on the right it would be easier to measure using the outer scale to get a measure of 129°.

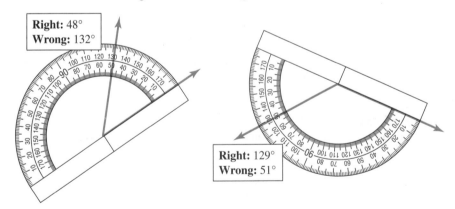

Right: 48°
Wrong: 132°

Right: 129°
Wrong: 51°

GOOD QUESTION:

Can an angle be more than 180 degrees? Can an angle be negative?

Answer: Yes, and you may have an intuition about what these types of angles would look like. These types of angles do exist and are used in other classes. Auto Mechanics 101, for example, would talk about an engine whose crankshaft turns 720° in a complete cycle. For this supplement, however, we will only explore angles that measure between 0° and 180°.

■ Notation for Angle Measure and Congruence

The angle in the margin is ∠*QRS*, or ∠*R*. The notation used to express its measure is *m*∠*QRS*, or *m*∠*R*. The reason for the distinction is that the angle itself and its measure are two different things (the same way that you and your height are two different things). You may feel the urge to write ∠*QRS* = 29° instead of *m*∠*QRS* = 29°. The exact level of rigor is ultimately up to your instructor, but this book will use the formal notation: *m*∠*QRS* = 29°.

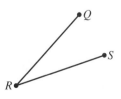

The same issue comes up when comparing angles. Recall from Section 1 that two geometric objects that have the same measure(s) are called **congruent.** Consider the following angles:

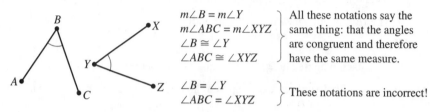

$m\angle B = m\angle Y$
$m\angle ABC = m\angle XYZ$
$\angle B \cong \angle Y$
$\angle ABC \cong \angle XYZ$

All these notations say the same thing: that the angles are congruent and therefore have the same measure.

$\angle B = \angle Y$
$\angle ABC = \angle XYZ$

These notations are incorrect!

Note that we use arcs on two angles to indicate they are congruent. This technique is similar to the tick marks used to indicate congruent line segments.

■ Common Relationships between Angles

In certain situations, relationships between angles are always true. There are four such situations we will discuss in this section, and several more will come up as we progress through this supplement.

Vertical Angles

Any time two lines intersect, four angles are created. The angles on opposite sides of the point of intersection are called **vertical angles.**

It can be proven that vertical angles have the same measure. That is, in the adjacent figure,

$$m\angle 1 = m\angle 3 \text{ and } m\angle 2 = m\angle 4$$
or
$$\angle 1 \cong \angle 3 \text{ and } \angle 2 \cong \angle 4$$

This is sometimes referred to as the **vertical angle theorem.** The proof of this theorem is explored in the Geometry Project at the end of this section.

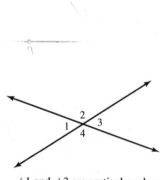

∠1 and ∠3 are vertical angles.
∠2 and ∠4 are vertical angles.

EXAMPLE 2–2   Using Algebra with Vertical Angles

Find the measure of the indicated angles in the figure.

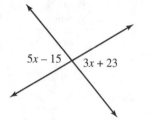

$5x - 15$ $3x + 23$

◆ SOLUTION

Using the vertical angle theorem leads us to the equation

$$5x - 15 = 3x + 23$$
$$2x = 38$$
$$x = 19$$

The actual measures are $5x - 15$ and $3x + 23$. So to find the angle measures we must evaluate these expressions: $5(19) - 15 = 80$ and $3(19) + 23 = 80$. So both angles are 80°. If these two values had come out to be different, that would have told us we had made an error in solving.   ◆

If two lines, segments, or rays intersect and one of the angles created is a right angle, it can be shown that all four of the angles formed are right angles and the lines are called **perpendicular.** If a perpendicular line, ray, or segment also bisects a line segment, then it is called a **perpendicular bisector.** In the figure in the margin, $\overline{JW}$ is a perpendicular bisector for $\overline{HS}$.

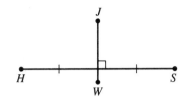

Adjacent Angles

Adjacent angles have the same vertex, share a side, and do not overlap. Notice the figure to the right. Here are some examples.

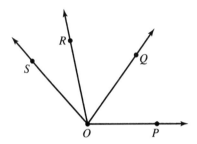

Adjacent angles:

   $\angle SOR$ and $\angle ROP$

   $\angle ROQ$ and $\angle QOP$

   $\angle SOQ$ and $\angle QOP$

Nonadjacent angles:

   $\angle SOR$ and $\angle QOP$ (do not share a side)

   $\angle SOQ$ and $\angle ROP$ (overlap)

   $\angle ROP$ and $\angle QOP$ (overlap)

Notice that $m\angle SOR + m\angle ROQ + m\angle QOP = m\angle SOP$. This *addition property of adjacent angles* should help give you an idea of why adjacent angles are not allowed to overlap. Notice, for example, that $m\angle QOP + m\angle ROP \neq m\angle ROP$, even though they share a side.

EXAMPLE 2-3   Using Algebra with Adjacent Angles

Suppose an angle that measures 164° is split into two adjacent angles. The measure of the larger angle is 112° less than five times the measure of the smaller angle. Find the measure of each angle.

◆ **SOLUTION**

We begin by drawing a figure to represent the situation, letting x represent the measure of the smaller angle.

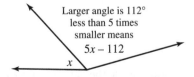

Larger angle is 112°
less than 5 times
smaller means
$5x - 112$

x

Since the angles are adjacent, by the addition property of adjacent angles, their combined measure must be 164°.

$$(x) + (5x - 112) = 164$$
$$6x = 276$$
$$x = 46$$

So the smaller angle is 46° and the larger is $5(46) - 112 = 118°$.

Finally, we can do a quick check to make sure the two answers add up to 164 degrees: $46° + 118° = 164°$. ✓   ◆

EXAMPLE 2-4   Using the Addition Property of Adjacent Angles

In the figure below, find $m\angle SHR$ given that $m\angle DHE = 100°$.

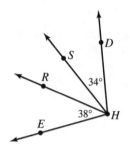

◆ **SOLUTION**

From the addition property of adjacent angles discussed earlier:

$$m\angle DHS + m\angle SHR + m\angle RHE = 100$$
$$34 + m\angle SHR + 38 = 100$$
$$m\angle SHR = 28$$ ◆

We mentioned in the previous section that to **bisect** an object means to divide it into two equal parts. In the case of an angle, an **angle bisector** is a ray that divides the angle into two congruent, adjacent angles. This next example makes use of this idea.

EXAMPLE 2-5   Using Information about a Bisector

In the figure below, the ray $\overrightarrow{AY}$ bisects $\angle TAL$. Find the measure of the two unlabeled angles if $m\angle KAL = 127°$.

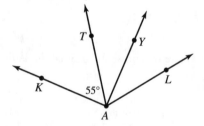

◆ **SOLUTION**

As in the last example, we want to use the addition property of adjacent angles. In this case there are two missing angle measures: $m\angle TAY$ and $m\angle YAL$. Since the ray $\overrightarrow{AY}$

bisects ∠*TAL*, the measures of these two unknown angles are equal, and we can use one variable (we will use *x*) to represent them both. This gives us the equation

$$m\angle KAT + m\angle TAY + m\angle YAL = 127$$
$$55 + x + x = 127$$
$$2x = 72$$
$$x = 36$$

So $m\angle TAY = m\angle YAL = 36°$. ◆

In the Construction Example at the end of this section, we will show how to bisect an angle without the use of a protractor.

Complementary and Supplementary Angles

Two angles are called **complementary** if their measures add up to 90°. Two angles are called **supplementary** if their measures add up to 180°.

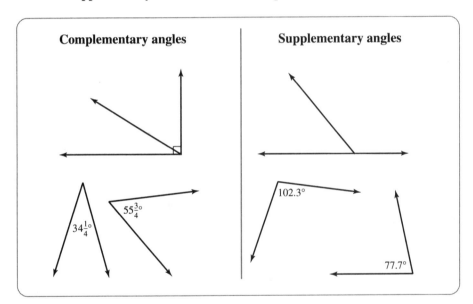

| **Complementary angles** | **Supplementary angles** |

As the figure indicates, the angles do not actually have to be adjacent to be complementary or supplementary.

The relationship between complementary and supplementary angles can be expressed algebraically as follows:

If ∠*A* and ∠*B* are complementary, then $m\angle A + m\angle B = 90°$.

If ∠*A* and ∠*B* are supplementary, then $m\angle A + m\angle B = 180°$.

EXAMPLE 2–6   Finding a Complement and Supplement

Find the complement and supplement of a 68.3° angle.

◆ **SOLUTION**

Let c = the measure of an angle complementary to 68.3°. Then

$$c + 68.3 = 90$$
$$c = 21.7$$

So any angle that measures 21.7° would be complementary to a 68.3° angle.

Let s = the measure of an angle supplementary to 68.3°. Then

$$s + 68.3 = 180$$
$$s = 111.7$$

So any angle that measures 111.7° would be supplementary to a 68.3° angle.   ◆

EXAMPLE 2-7   Using Algebra with Supplementary Angles

What angle is 48° more than three times its own supplement?

◆ **SOLUTION**

We begin by drawing a figure to help with organizing the information. Since the angles are supplementary, they can be drawn as adjacent angles along a straight angle.

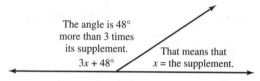

The angle is 48° more than 3 times its supplement.
$3x + 48°$

That means that x = the supplement.

We generate the equation

$$(3x + 48) + x = 180$$
$$4x = 132$$
$$x = 33$$

Remember, the variable x represents the supplement of the angle we are looking for, so to find the answer, we evaluate $3(33°) + 48° = 147°$. The solution is 147°. Finally, we do a quick check to make sure that 33° and 147° are supplementary: $33° + 147° = 180°$ ✓◆

EXAMPLE 2-8   Putting It All Together

The following figure has been labeled with angle measures.

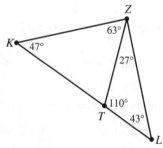

Use the figure to do the following exercises:

a. Give an obtuse angle.

b. Find the measure of $\angle KTZ$.

c. Give a pair of adjacent angles.

d. Give two supplementary angles.

e. Give two complementary angles that are adjacent.

f. Give two complementary angles that are not adjacent.

◆ **SOLUTION**

Success in a problem like this requires a clear understanding of the terminology and notation that have been addressed so far in this section.

a. There is only one obtuse angle in the figure: $\angle ZTL$.

b. Note that $\angle KTZ$ is adjacent to $\angle ZTL$, and together they form a straight angle, so

$$m\angle KTZ + m\angle ZTL = 180$$
$$m\angle KTZ + 110 = 180$$
$$m\angle KTZ = 70$$

c. We could give the pair $\angle KZT$ and $\angle TZL$ or we could give the pair $\angle KTZ$ and $\angle ZTL$ since both these pairs are adjacent.

d. The only supplementary angles in the picture are $\angle KTZ$ and $\angle ZTL$.

e. $\angle KZT$ and $\angle TZL$ are complementary since $m\angle KZT + m\angle TZL = 63° + 27° = 90°$. In addition, these angles are adjacent.

f. $\angle ZKT$ and $\angle TLZ$ are complementary since $m\angle ZKT + m\angle TLZ = 47° + 43° = 90°$. In addition, these angles are not adjacent.   ◆

■ Parallel Lines Cut by a Transversal

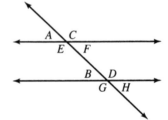

The figure in the margin shows two parallel lines both being intersected by a **transversal.** A transversal is just a line that intersects two or more other lines. The lines intersected by a transversal do not have to be parallel, but if they are, then there are simple, provable relationships between the angles created.

Notice there are eight angles in the figure. First off, the angles come in vertical pairs. There are other relationships between the angles. To discuss these relationships, there is some terminology we will need. We now summarize both the terms and the theorems.

Terminology

Interior angles lie on the interior of the two parallel lines. In our figure, angles E, F, B, and D are examples of interior angles. **Exterior angles** lie on the exterior of the two parallel lines (e.g., angles A, C, G, and H in the figure). **Same-side angles** lie on the same side of the transversal. Angles A, E, B, and G and angles C, F, D, and H in the figure are all same-side angles. **Alternate angles** lie on opposite sides of the transversal. Any two angles that are on opposite sides, such as angles G and F or angles E and D in the figure, are considered to be alternate angles. **Corresponding angles** lie on the same side of the transversal and in the same position with respect to the parallel lines (e.g., angles A and B, angles H and F, or angles E and G in the figure).

Theorems

The following theorems pertain to parallel lines being cut by a transversal:

• Corresponding angles are congruent.

• Alternate interior angles are congruent.

• Same-side interior angles are supplementary.

> **Congruent angles:**
> $\angle A \cong \angle F \cong \angle B \cong \angle H$
> $\angle C \cong \angle E \cong \angle D \cong \angle G$
>
> **Supplementary angles:**
> $m\angle E + m\angle B = 180°$
> $m\angle F + m\angle D = 180°$
> $m\angle A + m\angle G = 180°$
> $m\angle C + m\angle H = 180°$

- Same-side exterior angles are supplementary.
- Alternate exterior angles are congruent.

These theorems should seem intuitive. In the margin we have summarized all the important information from the theorems in a purely symbolic form.

EXAMPLE 2-9   Finding Angle Measures, Given Parallel Lines

In the figure below, lines n and m are parallel. Find the measure of every angle in the figure, given the measure of the indicated angle.

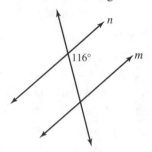

◆ SOLUTION

Using the theorems above, we find that all eight of the angles in the figure have one of two measures: 116° or 64°.

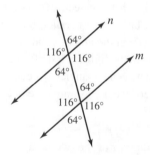

◆

EXAMPLE 2-10   Using Algebra with Parallel Lines

In the figure below you are given parallel lines cut by a transversal and algebraic expressions for two of the angles involved. Find the measure of the two angles.

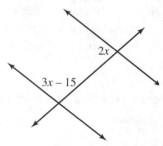

◆ SOLUTION

The angles given algebraic expressions are *same-side interior,* so they are supplementary:

$$(3x - 15) + (2x) = 180$$
$$5x - 15 = 180$$
$$5x = 195$$
$$x = 39$$

So the angles measure $2(39) = 78°$ and $3(39°) - 15 = 102°$.   ◆

CONSTRUCTION EXAMPLE

Using a Straightedge and Compass to Bisect an Angle

Let's revisit the idea of constructing in geometry without the use of a measuring tool. In the first section, we showed how to bisect a line segment using only a straightedge and compass. In this section, we will look at how to bisect an angle. Recall that an angle bisector is a ray that divides the angle into two smaller angles with equal measure. Here is the procedure used to accomplish this using only a straightedge and compass:

1. Begin with any angle.

2. Open the compass to any distance and make an arc that crosses both rays of the angle. Mark the points where the arc crosses the rays.

3. Now decrease the distance of the compass and make an arc in the interior of the angle by starting at one of the points where the first arc crosses a ray.

4. Now, keeping the compass at the same distance, repeat the previous step for the other point of intersection. Mark the points where the arcs from steps 3 and 4 intersect.

5. The vertex of the angle and the points where the two arcs intersect are collinear points. In addition, the ray created by these points is a bisector for the original angle.

As with the exercise on bisecting a line segment in Section 1, you will have a few exercises that involve this type of construction.

A similar type of problem that the Greeks (and many after them) attempted was the challenge of *trisecting* an angle (i.e., dividing it into three congruent angles).

Trisection

Mathematicians tried for hundreds of years to accomplish this using only a compass. It wasn't until the 1800s that it was finally proven using algebra that it's an impossible challenge; it can't be done.

Section 2 | Exercises

*For problems 1–6, use the terms from the **Vocabulary Checklist** to fill in the blanks.*

✓

VOCABULARY CHECKLIST:

acute angle	*adjacent angles*
right angle	*complementary angles*
obtuse angle	*supplementary angles*
straight angle	*transversal*
protractor	*interior angles*
vertical angles	*exterior angles*
perpendicular	*same-side angles*
perpendicular bisector	*corresponding angles*

1. Two lines intersecting always creates _____ angles, _____ angles, and _____ angles.

2. If two angles are complementary, they must both be _____ angles.

3. Any line that intersects two other lines is called a(n) _____.

4. If two angles are supplementary and one of them is an acute angle, then the other angle must be a(n) _____ angle.

5. _____ lines will create four right angles.

6. _____ angles are angles that lie on the same side of a transversal.

For exercises 7–10, answer TRUE or FALSE.

T | F   7. Two angles whose measures add to 180° are called complementary.

T | F   8. Corresponding angles may lie on different sides of a transversal.

T | F   9. Any two angles that share an endpoint are adjacent.

T | F 10. Vertical angles are always supplementary.

11. **Astronomy.** The following figure gives an example of planetary alignment.

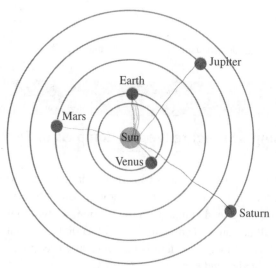

Using the figure as a guide, and the sun as the vertex, classify the angle between the Earth and each other planet as acute, obtuse, or approximately right.

12. On a directional compass, how many degrees are between North and East?

13. On a directional compass, how many degrees are between North and South?

14. On a directional compass, how many degrees are between South and South-East?

15. On a directional compass, how many degrees are between East and South-West?

16. On a clock, how many degrees are between hour 12 and hour 3?

17. On a clock, how many degrees are between hour 2 and hour 6?

18. On a clock, how many degrees are between minute 5 and minute 35?

19. On a clock, how many degrees are between minute 13 and minute 28?

20. Name every angle in the figure below and classify each as acute, right, or obtuse.

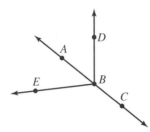

21. Name every angle in the following figure and classify each as acute, right, or obtuse. Assume the angles are drawn to scale.

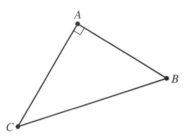

22. Name every angle in the figure below and classify each as acute, right, or obtuse.

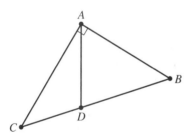

23. Draw three different figures that show a line segment and a ray intersecting to form:

    a. four angles

    b. one angle

    c. two angles

24. Is it possible to have two lines intersect that form four acute angles? Draw a figure to help explain.

25. Draw the following rays so that the information is true:
    Rays: $\overrightarrow{AB}, \overrightarrow{AC}, \overrightarrow{AD}$

    Information: $\angle CAB$ is acute; $\angle DAB$ is a straight angle.

26. Draw the following rays so that the given information is true:

    Rays: $\overrightarrow{PG}, \overrightarrow{PK}, \overrightarrow{PT}, \overrightarrow{PH}$

    Information: $\angle GPK$ is acute; $\angle KPT$ is obtuse; $\angle TPH$ is a right angle.

27. Use a protractor to measure each angle and classify it as acute, obtuse, or right.

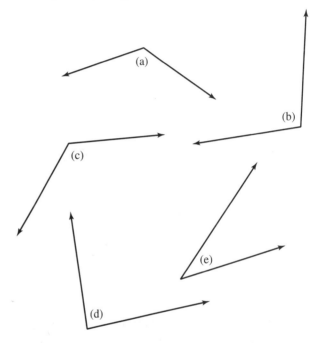

Exercises 28–31 refer to the following figure.

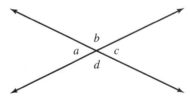

28. Which angles are vertical angles?

29. Which angles are supplementary angles?

30. Find $m\angle a$, $m\angle b$, and $m\angle c$, given that $m\angle d = 130°$.

31. Find the measure of all four angles, given that the measure of $\angle b$ is twice the measure of $\angle a$.

32. Is it possible for two angles to be vertical angles *and* complementary angles? What about vertical *and* supplementary? Explain.

33. In each figure, find *x*.

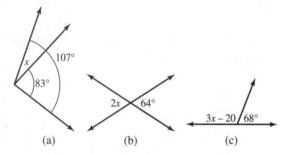

    (a)   (b)   (c)

34. In each figure, find *x*.

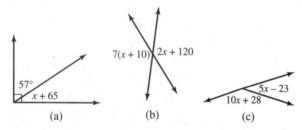

    (a)   (b)   (c)

35. *Fill in the blanks:* If an angle measures 148°, then an angle bisector will divide the angle into _____ angles that each measure _____.

36. *Fill in the blanks:* If an angle measures 57°, then an angle bisector will divide the angle into _____ angles that each measure _____.

37. *Fill in the blanks:* If an angle measures 106.8°, then an angle bisector will divide the angle into _____ angles that each measure _____.

38. *Fill in the blanks:* If an angle measures $78\frac{3}{4}°$, then an angle bisector will divide the angle into _____ angles that each measure _____.

39. Classify each pair of angles as complementary, supplementary, or neither.

   a. 129° and 41°

   b. 34° and 56°

   c. $67\frac{3}{5}°$ and $112\frac{4}{10}°$

   d. 45.0034° and 44.9965°

40. Find the complement of a 39° angle.

41. Find the supplement of a 39° angle.

42. Find the supplement of a 112° angle.

43. Find the complement of an 80.4° angle.

44. Find the complement of a 4.15° angle.

45. Find the supplement of a $68\frac{1}{3}°$ angle.

46. What angle is five times as large as its own complement?

47. What angle is seven times as large as its own supplement?

48. What angle is 10° more than three times its own complement?

49. What angle is 24° less than twice its own supplement?

50. Explain why the supplement of an obtuse angle cannot also be obtuse.

51. Explain why the supplement of an acute angle cannot also be acute.

Exercises 52–61 refer to the following figure, in which each angle has been measured and labeled.

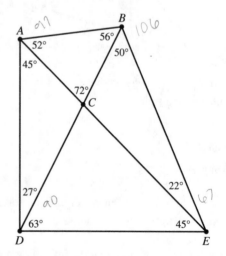

52. Name a right angle.

53. Name four obtuse angles.

54. Name the three angles whose measure has not been included and find their missing measures.

55. Name three pairs of adjacent angles.

56. Name two pairs of vertical angles.

57. Name two complementary angles that are adjacent.

58. Name two complementary angles that are not adjacent, if possible.

59. Name two supplementary angles that are adjacent.

60. Name two angles that are supplementary, but not adjacent, if possible.

61. Name two perpendicular line segments in the figure.

62. Using the figure below, solve for the value of x, given that $m\angle IES = 88°$.

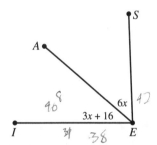

63. Using the figure below, solve for the value of x, given that $m\angle SOP = 156°$.

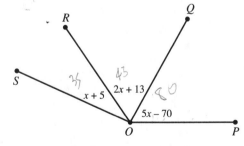

64. Draw an example of a transversal cutting two nonparallel lines.

65. Draw an example of a transversal cutting two intersecting lines.

66. Is it possible to draw a figure of a transversal cutting two perpendicular lines? If so, draw it.

67. Given that the lines are parallel, find the measure of every angle in the figure.

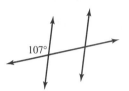

68. Given that the lines are parallel, find the measure of every angle in the figure below.

69. In the figure below, you are given expressions for two angles. Find the measure of those angles and the measure of every other angle in the figure.

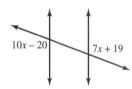

70. **Physics.** A light ray that bounces off a mirror will reflect at the same angle at which it hits. Suppose a light ray hits a mirror at a 23° angle. Draw a figure that represents the result. Find and label the measure of every angle in the figure.

For exercises 71–76 consider the following letters from our alphabet. If necessary, imagine the segments extended into lines.

A H F N T E X Z

71. Which letter(s) contain parallel line segments?

72. Which letter(s) contain perpendicular line segments?

73. Which letter(s) contain corresponding angles?

74. Which letter(s) contain vertical angles?

75. Which letter(s) contain a transversal?

76. Which letter(s) contain supplementary angles?

CONSTRUCTION

77. **Straightedge and Protractor.** Use a ruler and a protractor to construct an angle that is *exactly* 98°. Now divide the angle into three adjacent angles that measure 23°, 30°, and 45°.

78. **Straightedge, Compass, and Protractor.** Use a ruler to draw any acute angle. Now bisect the angle using only a compass, as outlined in the construction example in this section. Use a protractor to help verify that you have succeeded in bisecting the angle.

79. **Straightedge, Compass, and Protractor.** Use a compass and the techniques from the Construction Example to show that in the following figure, ray $\overrightarrow{BD}$ does not bisect angle *ABC*.

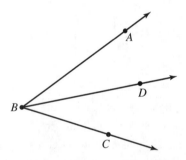

80. **Protractor, Straightedge, and Compass.** Use a protractor to create an angle that is exactly 45°. Now use the steps from the previous exercise to copy this angle. Use a protractor to verify that the angle you created is exactly 45°. If it is a little bit off, don't worry, that's human error.

81. **Straightedge and Compass.** Use a straightedge to draw any angle. Call it ∠*A*; that is, label its vertex point *A*. Now use these steps to make an exact copy of it.

1. Put the tip of the compass at point *A* and draw an arc that crosses both rays. Leave the compass at this width, as you will need it later.

2. Draw another ray. Call its endpoint *B*.

3. Keeping the compass at the same width as step 1, put the tip on point B and draw a wide arc that crosses the ray and extends wider than the angle you are copying.

4. Back to ∠*A*. Open the compass to the width between the two points where the arc intersects the rays.

5. Back to the other ray. Put the tip of the compass on the point where the arc crosses the ray and make a second arc that intersects the first arc made on this ray.

6. Make a second ray that has *B* as its endpoint and goes through the point where the two arcs intersect.

7. If this is done correctly, ∠*A* and ∠*B* will be congruent.

Problem 81 is a very good example of the kind of challenge the Greeks liked: Can you copy an angle without actually measuring it first? These steps show that you can.

Geometry Project 2 | Proofs in Geometry

This project explores how things are proven in geometry. We begin by exploring the difference between an axiom (or postulate) and a theorem.

An **axiom** (or **postulate**) is an assumption that can be made from basic information without requiring proof. One example of this is the parallel lines postulate.

> **Parallel Lines Postulate:** *Suppose two lines are cut by a transversal. The two lines are parallel if and only if the corresponding angles formed are congruent.*

To illustrate, look at the figure below.

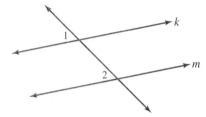

The postulate is saying that lines *k* and *m* are parallel if and only if angles 1 and 2 are congruent. The phrase "if and only if" means the converse is also true. So if angles 1 and 2 are congruent, then the lines are parallel, and if the lines are parallel, then angles 1 and 2 are congruent.

A **theorem** is a statement that can be proven using one or more postulates along with common math techniques. Here is an example of a theorem that is a direct result of the parallel lines postulate.

> **Alternate Interior Angles Theorem:** *If two parallel lines are cut by a transversal, then the alternate interior angles formed are congruent.*

In order to be accepted, theorems must be proven. The most common technique for proving theorems in geometry is to use a table that contains two columns: the first column holds a series of statements that leads to the proof of the theorem, and the second column holds the justification for each step. The justification may be a definition, postulate, or some property from mathematics that we know to be true. In this way we can ensure that all the steps are correct and therefore the theorem is indeed true.

You will be presented with the definitions and concepts used for justifying the geometry statements you make. Here are some useful properties from arithmetic and algebra that may be used as justification for steps in proofs.

The transitive property:
- If $a = b$ and $b = c$ then $a = c$.
- This property can also be applied to inequalities.

The addition, subtraction, or multiplication property of equations:
- If $a = b$ then $a + c = b + c$, $a - c = b - c$, and $a \cdot c = b \cdot c$.
- This property can also be applied to inequalities.

The substitution property:
- If $a = b$ then a can replace b in any equation or inequality.

Let's look at a two-column proof for the alternate interior angles theorem just mentioned.

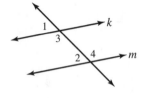

Prove: $\angle 3 \cong \angle 4$, given that lines k and m are parallel.

Statement	Justification
$m\angle 1 + m\angle 3 = 180°$ $m\angle 2 + m\angle 4 = 180°$	Definition of supplementary angles
$m\angle 3 = 180 - m\angle 1$ $m\angle 4 = 180 - m\angle 2$	The subtraction property of equations
$m\angle 1 = m\angle 2$	The parallel lines postulate
$180 - m\angle 1 = 180 - m\angle 2$	The subtraction property of equations
$m\angle 3 = m\angle 4$	Substitution ∎

The symbol ∎ means we are finished with the proof. Notice that while the statements we made are fairly intuitive we still had to put justification for each one.

Project Exercises | Writing Geometry Proofs

1. **Theorem:** *Vertical angles are congruent.* The following is an outline of the proof for this theorem. For each statement, fill in a justification. (*Hint:* One of the justifications is the transitive property of equality.)

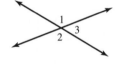

Prove: $m\angle 1 = m\angle 2$

Statement	Justification
$m\angle 1 + m\angle 3 = 180°$ $m\angle 2 + m\angle 3 = 180°$	?
$m\angle 1 = 180 - m\angle 3$ $m\angle 2 = 180 - m\angle 3$	?
$m\angle 1 = m\angle 2$	? ∎

2. **Theorem:** *If two parallel lines are cut by a transversal, then the alternate exterior angles formed are congruent.* Write a two-column proof for this theorem. The proof will be similar to the one provided above for the congruence of alternate interior angles.

3. **Theorem:** *If two parallel lines are cut by a transversal, then the same-side interior angles formed are supplementary.* Write a two-column proof for this theorem.

Triangles

A **triangle** is made up of three line segments that share endpoints. Given any three points A, B, and C that are noncollinear, we can form a triangle by creating the line segments $\overline{AB}$, $\overline{BC}$, and $\overline{AC}$.

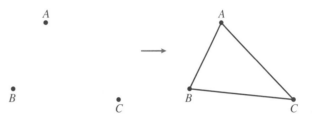

The points A, B, and C are the **vertices** (plural of vertex) of the triangle. This triangle could be named six ways using the symbol $\triangle$ and the three points that make up the vertices: $\triangle ABC$, $\triangle CAB$, $\triangle BCA$, and so on. So long as all three points are included, the notation refers to the same figure. Note the triangle has three angles: $\angle A$, $\angle B$, and $\angle C$. The side not involved in a given angle is said to be **opposite** that angle. So, for example, $\overline{BC}$ is opposite $\angle A$, and $\overline{AC}$ is opposite $\angle B$.

■ The Practical Nature of Triangles

How do astronomers find the distance between stars? How did the Greeks find the radius of the Earth over 2000 years before the advent of advanced technology? How can a surveyor find the height of a mountain that is miles away or the distance across a large body of water? How can a soldier find the distance between a cannon and its target?

The answer to all these questions is the same . . . by using triangles. The practical techniques that can be extracted from the study of triangles are so powerful that, by the end of Section 4: More about Triangles, you and a friend will be able to go downtown and find the height of any building you see using only household items. And how will you do it? Triangles.

Of course, in order to apply triangles at that level we must understand quite a bit about relationships that exist in all triangles. This section presents basic terminology for triangles and explores a few important features that all triangles share. We will also look at the famous Pythagorean Theorem, which pertains to triangles that have a right angle.

We begin in this section with some terminology involved with classifying triangles by their sides and angles.

■ Classifying Triangles

Triangles can be classified by their angles and/or their sides. The following table summarizes the terminology and significance of classifying triangles. A considerable amount of geometry involves work with triangles, so will use this terminology throughout the rest of the book.

CLASSIFICATION BY ANGLES

CLASSIFICATION	DESCRIPTION	EXAMPLE
Acute triangle	All three angles in the triangle are less than 90°.	
Right triangle	One of the angles in the triangle is a right angle. The side opposite the right angle is called the **hypotenuse.** The sides that make up the right angles are called **legs.**	
Obtuse triangle	One of the angles in the triangle is between 90° and 180°.	

CLASSIFICATION BY SIDES

CLASSIFICATION	DESCRIPTION	EXAMPLE
Scalene triangle	All three sides of the triangle have different lengths. Note the use of tick marks to indicate the sides are different. All angles have different measures.	
Isosceles triangle	Two of the sides of the triangle have the same length. Note the use of tick marks to indicate the congruent sides. As indicated, the angles opposite the congruent sides are also congruent.	
Equilateral triangle	All three sides of the triangle have the same length. Note the use of tick marks to indicate that all three sides are congruent. In an equilateral triangle, all three angles are congruent as well.	

EXAMPLE 3–1   Sketching Triangles

Sketch an example of an isosceles obtuse triangle and a scalene right triangle.

◆ **SOLUTION**

For an isosceles obtuse, we make sure that two sides are congruent and that one of the angles is obtuse. For the scalene right, we make sure that all sides have different lengths and that the triangle has one right angle.

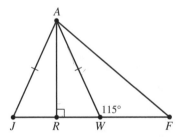

isosceles obtuse   scalene right   ◆

EXAMPLE 3–2   Finding and Classifying Triangles

In the following figure, the only congruent sides are the ones indicated and the angles are drawn to scale. Name every triangle shown and classify it by its angles and sides, if possible.

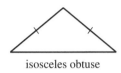

◆ **SOLUTION**

There are six triangles total. All but one can be classified using the information given:

- △AJR is a right scalene triangle since it contains a 90° angle and none of its sides are marked as congruent.

- △ARW is also a right scalene triangle.

- △AWF is an obtuse scalene triangle since it contains a 115° angle and none of its sides are marked as congruent.

- △AJW is an acute isosceles triangle since two of its sides are congruent. We were not given the angle measures, but since we were told the angles are drawn to scale we can see they are all acute.

- △ARF is another right scalene triangle.

- △AJF cannot be classified by its angles. ∠JAF looks like a right angle, but since it isn't labeled as right it could be slightly acute or obtuse. We can, however, classify △AJF as a scalene triangle, since its sides are not marked as congruent.   ◆

EXAMPLE 3-3   Using a Ruler and Protractor

Consider the triangle below. Use a ruler and a protractor to help with the following exercises. Measure each side in centimeters, measure each angle.

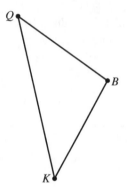

a. Classify the triangle by its angles.
b. Classify the triangle by its sides.
c. What angle is opposite side $\overline{BK}$?

◆ **SOLUTION**

Below is an image of a protractor being used to measure an angle of the triangle and an image of the triangle after being measured.

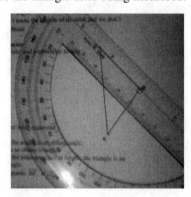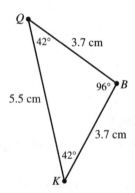

a. Since one of the angles is an obtuse angle, the triangle is an obtuse triangle.
b. Since two of the sides are equal in length, the triangle is an isosceles triangle.
c. The angle opposite $\overline{BK}$ is $\angle Q$.

The classification of this triangle involves both its sides and its angles, so we would call it an obtuse isosceles triangle.   ◆

EXAMPLE 3-4   The Medians and Angle Bisectors of a Triangle

A **median** of a triangle is a line segment drawn from one vertex to the opposite side that bisects the side. An **angle bisector** is a line segment drawn from one vertex to the opposite side that bisects the corresponding angle. These objects are popular in theorems and proofs in geometry, but in this example we just want to become familiar with their definition and how they are drawn. Consider the following triangle.

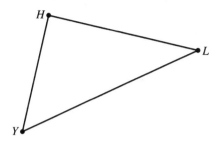

Sketch and label the features involved with an angle bisector from point H and a median from point Y.

◆ **SOLUTION**

From point H we need a line segment that bisects $\angle H$ and from point Y we need a line segment that bisects line segment $\overline{HL}$. Here is the result.

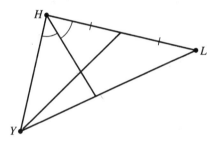

Note the use of tick marks and angle arcs to indicate which angles and line segments are congruent.   ◆

■ The Angles of a Triangle

Interior Angles

The sum of the three angles in a triangle is always 180 degrees. You can convince yourself of this by cutting any triangle out of any piece of paper, tearing the corners off, and setting them up as adjacent angles. This will always result in a line. And since we know that a line segment has an angle measure of 180 degrees, then we have the following.

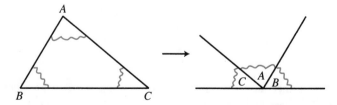

$$m\angle A + m\angle B + m\angle C = 180°$$

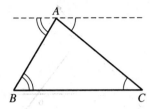

This is very easy to prove. By drawing a line through point *A* that is parallel to line segment $\overline{BC}$, we have the figure in the margin.

Notice the three adjacent angles that form a straight angle and the congruent angles made by the parallel lines. This can be used to show that $m\angle A + m\angle B + m\angle C = 180°$. We will present a more formal proof in the Geometry Project at the end of this section.

As we have seen in other sections, this idea leads to an algebraic equation when angles from a triangle are involved in a problem.

EXAMPLE 3-5   Finding an Angle Measure

If two angles in a triangle measure $54\frac{4}{9}°$ and $87\frac{5}{12}°$, what is the measure of the third angle?

♦ **SOLUTION**

Let *C* represent the measure of the third angle. From the equation above we have

$$54\frac{4}{9} + 87\frac{5}{12} + C = 180$$

$$141\frac{31}{36} + C = 180$$

$$C = 180 - 141\frac{31}{36} = 38\frac{5}{36}$$

Therefore, the third angle is a $38\frac{5}{36}°$ angle.   ♦

EXAMPLE 3-6   Using Algebra with the Interior Angles in a Triangle

The second angle in a triangle is twice the first. The third is three times as large as the second. What is the measure of each angle in the triangle?

♦ **SOLUTION**

We first organize the information in a figure.

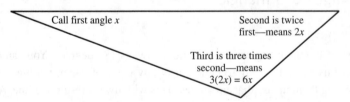

Since these three angles come from the same triangle, we have the equation

$$x + (2x) + 3(2x) = 180$$
$$x + 2x + 6x = 180$$
$$9x = 180$$
$$x = 20$$

So the measure of the first angle is 20 degrees. The measure of the second angle is $2(20°) = 40°$. The measure of the third angle is $3(40°) = 120°$. A quick check verifies the angles add up to 180°: $20° + 40° + 120° = 180°$.✓   ♦

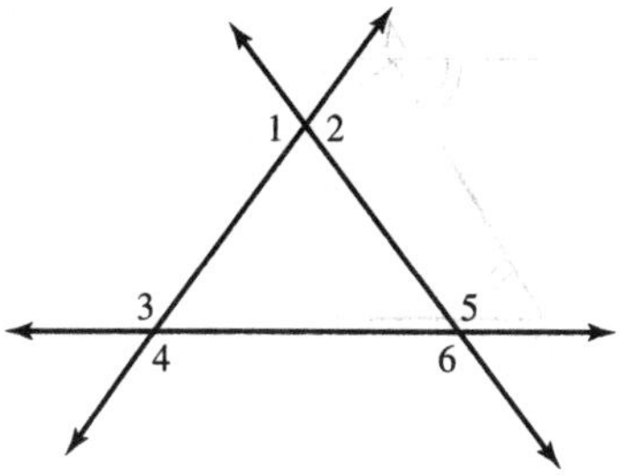

Exterior Angles

If the line side of a triangle is extended into a ray or line, an **exterior angle** is created. The figure in the margin illustrates that there are six such angles that can be associated with every triangle.

Notice that every exterior angle is supplementary to the corresponding interior angle. This simple idea can be used to prove the following theorems about exterior angles.

- An exterior angle of a triangle has a degree measure equal to the sum of the measures of the two interior angles that do not correspond to that exterior angle.
- An exterior angle of a triangle will always have a greater measure than either of the interior angles that do not correspond to it.

EXAMPLE 3-7 Finding Angle Measures

Sketch $\triangle ABC$ and label the measure of each of its interior and exterior angles, given that $m\angle A = 87°$ and that the exterior angles corresponding to $\angle C$ measure 120°.

◆ **SOLUTION**

We make an initial sketch of the triangle to show what measures we have been given. Notice that we also include the exterior angle in the figure.

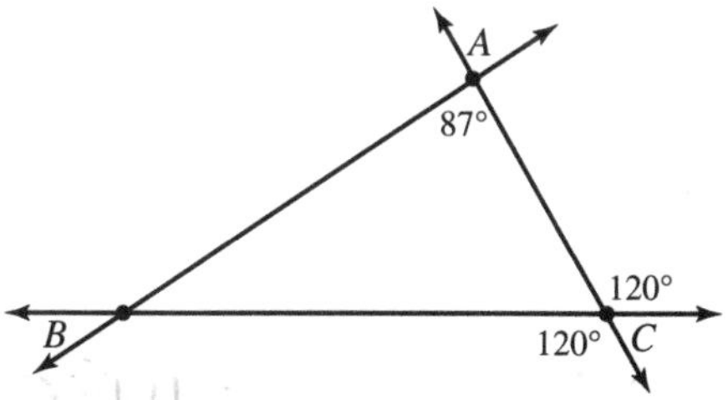

So we need to find $m\angle C$, $m\angle B$, and the measure of all the exterior angles. In order to do this, we take the following steps:

1. $\angle C$ is supplementary to its exterior angles, so $m\angle C = 180 - 120 = 60°$.
2. The sum of the interior angles must be 180°, so $m\angle B = 180 - 87 - 60 = 33°$.
3. Each exterior angle is supplementary to its corresponding interior angle, so their measures must be
   - $180 - 87 = 93°$
   - $180 - 33 = 147°$
4. We could also note that since the exterior angles come in vertical pairs, the two corresponding to the same interior angle are congruent.

Here is the final result.

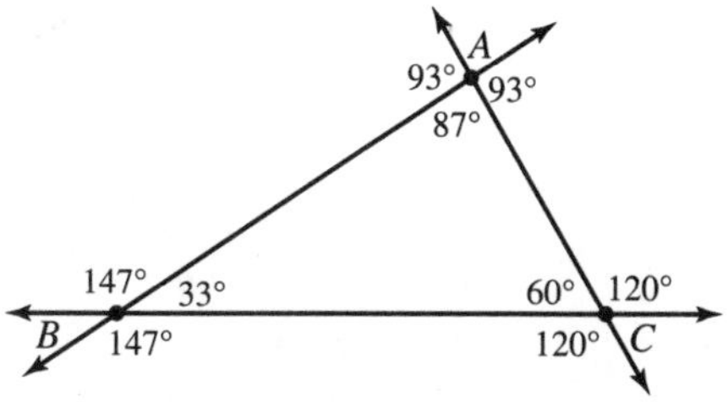

◆

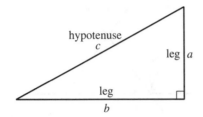

Right Triangles: The Pythagorean Theorem

In the figure in the margin, we have given some terminology associated with right triangles. The **hypotenuse** is the side opposite the right angle, and the **legs** make up the right angle. In addition, the variables a, b, and c have been used to represent the lengths of the sides of a right triangle. The relationship between the sides is that the sum of the squares of the legs is equal to the square of the hypotenuse, or $a^2 + b^2 = c^2$. This is called the **Pythagorean Theorem,** attributing its discovery to the Greek philosopher Pythagoras about 2500 years ago. This idea was explored by ancient civilizations that predate the Greeks, but Pythagoras was the first to provide proof.

There are dozens of ways to prove the Pythagorean Theorem, though not all the proofs are easy. But, in fact, U.S. President James Garfield even came up with a fairly simple way of proving the theorem.

EXAMPLE 3-8   Using the Pythagorean Theorem

Find the hypotenuse of a right triangle if $a = 6$ ft and $b = 8$ ft.

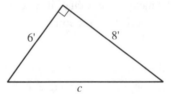

◆ **SOLUTION**

By the Pythagorean Theorem $c^2 = a^2 + b^2$, so $c^2 = 6^2 + 8^2 = 36 + 64 = 100$, so $c = 10$ feet.

Notice that it is *not* true that $6 + 8 = 10$. That is, the relationship involves the square of the side lengths, not just the side lengths themselves.   ◆

There are certain combinations of numbers, called **Pythagorean triples,** for which the lengths of all three sides of a right triangle are whole numbers. Some examples of Pythagorean triples are 3-4-5, 5-12-13, and 38-80-89; the two smaller numbers being the lengths of the legs and the largest the length of the hypotenuse. Usually, however, finding the length of an unknown side will involve either leaving your answer in radical form (exact form) or approximating the square root with a calculator.

EXAMPLE 3-9   Using the Pythagorean Theorem and Approximating

Find b.

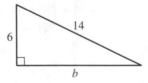

◆ **SOLUTION**

$$6^2 + b^2 = 14^2$$
$$36 + b^2 = 196$$
$$b^2 = 196 - 36 = 160$$
$$b = \sqrt{160} \approx 12.6$$

◆

Certain right triangles with special acute angles have other important properties. We will explore them in Section 4.

Comparing the Sides and Angles of a Triangle

While the angles of a triangle are not proportional to the lengths of the sides opposite them, there are important and useful relationships between the angles and sides.

- The largest side of a triangle is the side opposite the largest angle.
- The smallest side is the side opposite the smallest angle.
- As mentioned before, if two angles have the same measure, then the triangle is isosceles. That is, the sides opposite those angles will be congruent.
- The sum of the lengths of any two sides of a triangle must be greater than the length of the third side.

EXAMPLE 3–10   What's Wrong with This Picture?

Name three things that show that the triangle below could not exist.

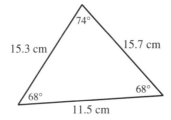

◆ SOLUTION

1. The angles do not add up to 180°. This can't be correct!

2. The side opposite the largest angle should be the largest side. So the side that is 11.5 cm should be the longest, but it is labeled as the shortest. This can't be correct!

3. The triangle has two equal angles. The sides opposite those angles should have exactly the same length, but they do not. This can't be correct!   ◆

CONSTRUCTION EXAMPLE

Using a Compass to Create an Equilateral Triangle

Once again, let's take a look at what the Greeks considered to be true geometry. In this case, we will create an equilateral triangle using nothing but a compass and a straightedge. This would be an easy task if we used a protractor. But to the Greeks, that would be cheating! While this may seem like quite a challenge, it's actually easier than the one in the last section. Here is the procedure:

1. Begin by creating a line segment that represents the length of one side of your equilateral triangle.

2. Open the compass to the length of this line segment and make an arc with the point of the compass at one of the endpoints of the line segment.

3. Repeat step 2 for the other endpoint and mark where the two arcs intersect.

4. This point of intersection is the missing vertex for your triangle. Simply use this point and the endpoints of the original line segment to create the remaining two sides.

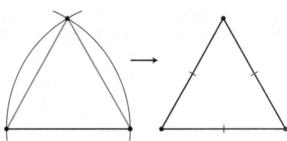

As before, the homework contains an exercise that allows you to try this for yourself.

Section 3 | Exercises

For exercises 1–6, use the terms from the **Vocabulary Checklist** *to fill in the blanks.*

✔

VOCABULARY CHECKLIST:

vertex (vertices)

acute triangle

right triangle

hypotenuse

leg

obtuse triangle

scalene triangle

isosceles triangle

equilateral triangle

Pythagorean Theorem

Pythagorean triple

1. If exactly two of the sides in a triangle are congruent, then the triangle is a(n) _____.

2. A triangle with an angle that measures 118° is considered a(n) _____.

3. If the term *hypotenuse* is used for a triangle, we know that the triangle is a(n) _____.

4. With regard to a right triangle, the numbers 3, 4, 5 are called a(n) _____.

5. The two legs of a right triangle form the angle that is opposite the _____ of the right triangle.

6. The three _____ of a triangle are the endpoints of the line segments that form the triangle.

For exercises 7–10, answer TRUE or FALSE.

T | F   7. The Pythagorean Theorem is only relevant for right triangles.

T | F   8. In an obtuse triangle, the side opposite the obtuse angle may or may not be the longest side.

T | F   9. Every triangle has exactly three vertices.

T | F 10. An acute triangle can never be isosceles.

Exercises 11–18 refer to the triangle below.

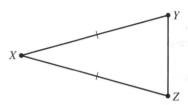

11. Name the triangle.

12. What are the vertices of the triangle?

13. Name the sides of the triangle.

14. Name the angles of the triangle.

15. Classify the triangle according to its sides.

16. Classify the triangle according to its angles.

17. What angle is made by the line segments $\overline{YZ}$ and $\overline{XZ}$?

18. What angle is opposite the side $\overline{XZ}$?

The figure below shows the most common way in geometry to represent a generic triangle. The angles are labeled with uppercase letters and the sides opposite each angle are labeled with the same letter in lowercase. Then values of the angles and/or sides can be given in order to explore a specific problem. One convention with this scenario is that the triangle may not be drawn to scale. We may say, for example, in one problem that $m\angle B = 103°$ even though in the figure $\angle B$ does not look obtuse. Keep that in mind.

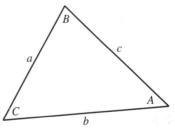

Exercises 19–26 refer to the preceding figure. In each case, some information is given about our triangle. Use the information to classify the triangle by its sides and/or its angles, whichever is possible.

19. $a = 5, b = 9, c = 7$

20. $a = 3, c = 3, B = 64°$

21. $b = 4, c = 6, A = 125°$

22. $a = 19.2, b = 19.2, C = 60°$

23. $b = 5.7, c = 2.9, A = 90°$

24. $A = 54°, B = 81°$

25. $a = 7.9, B = 90°, c = 7.9$

26. $C = 109°$

27. Which of the following is impossible to draw (there may be more than one)?

    a. a scalene obtuse triangle

    b. an equilateral right triangle

    c. an isosceles scalene triangle

    d. an isosceles acute triangle

28. What type of triangle would be created by a person on third base, a person on first base, and a person at home plate?

29. What type of triangle would be created by a surveyor, a point at the bottom of a mountain, and the peak of the mountain?

30. A line from the top of a tower to a point on the ground near the tower is called a *guy wire*. Guy wires are used to help support towers. What type of triangle would be created by a guy wire supporting a telephone pole?

31. What type of triangle would be created by the trusses of an A-frame house?

32. What type of triangle would be created by an astronomer looking at endpoints on the sun or moon?

33. Name every triangle in the figure below and classify each of them using the information given, if possible. Assume the angles are drawn to scale.

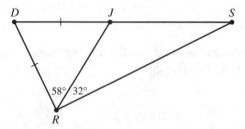

34. Name every triangle in the figure below and classify each of them using the information given, if possible.

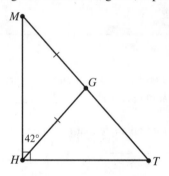

35. In the following figure draw and label features for a median from point S and an angle bisector from point H.

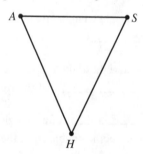

36. The measures of two angles in a triangle are 78° and 34°. What is the measure of the third angle?

37. The measures of two angles in a triangle are 100° and 70°. What is the measure of the third angle?

38. The measures of two angles in a triangle are 30° and 60°. What is the measure of the third angle?

39. The measures of two angles in a triangle are 45° and 90°. What is the measure of the third angle?

40. The measures of two angles in a triangle are 19.6° and 83.5°. What is the measure of the third angle?

41. The measures of two angles in a triangle are $54\frac{1}{6}°$ and $65\frac{3}{10}°$. What is the measure of the third angle?

42. In an equilateral triangle, all three angles are congruent. If this is the case, what is the measure of each angle?

43. The first angle in a triangle is 35°. The second angle is 18° less than three times the first angle. The third angle is 23° more than the first angle. Find the measure of each angle in the triangle.

44. The second angle in a triangle is one and a half times the measure of the first. The third is 15° more than half the first angle. Find the measure of each of the angles in the triangle.

45. Find the measures of the three angles in an isosceles triangle if the noncongruent angle is three times as big as the two congruent angles.

46. Find the measure of the two congruent angles in an isosceles triangle if the noncongruent angle is 126°.

47. In the figure below, find x.

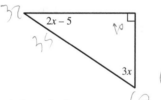

48. In the figure below, find x.

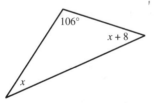

49. In the figure below, find x.

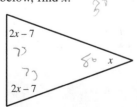

50. In the figure below, find $m\angle A$ and $m\angle ACB$, given that $m\angle ACD = 100°$ and $m\angle B = 35°$.

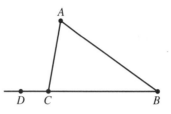

51. In the figure below, find $m\angle B$ and $m\angle BLS$, given that $m\angle A = 75°$ and $m\angle ALB = 30°$.

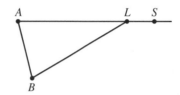

52. In the figure below, six angles are numbered. Find the measure of the other four angles if $m\angle 1 = 98°$ and $m\angle 4 = 135°$.

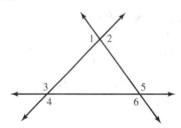

53. Draw a triangle $\triangle KGM$ and label each of its interior and exterior angles if $m\angle G = 50°$ and the exterior angles corresponding to $\angle M$ measure 100°.

54. Draw a triangle $\triangle SCA$ and label each of its interior and exterior angles if $m\angle S = 75°$ and the exterior angles corresponding to $\angle C$ measure 90°.

In exercises 55–60 use the Pythagorean Theorem to find the length of the missing side. Leave your answers in exact form, then approximate, if necessary.

55.

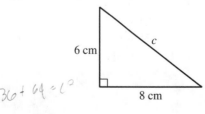

56.

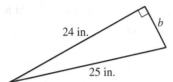

57.

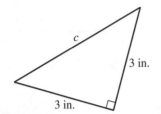

58.

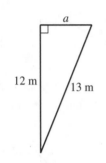

59.

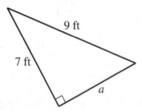

60.

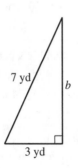

61. **Supporting a Tower.** Suppose a guy wire needs to be placed to support a tower that is 95 feet high. The wire will be secured to a spot 25 feet from the base of the tower. How long will the guy wire be?

62. **Surveying.** One classic surveyor challenge is to find the distance across a body of water. But this is easier than it may seem. A surveyor measures the distance between a third point that creates a right triangle, as shown in the figure below. What is the distance across the lake?

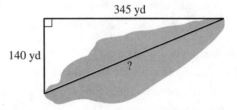

63. **Television Screen.** A television's inch measurement actually refers to the diagonal distance between corners. Assuming that the sides of the TV are equal in length, how many inches are the sides of a 19-inch TV?

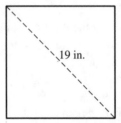

64. **The Wisdom of Homer.** *The Simpsons* ran an episode where Homer puts on a pair of glasses and takes a shot at the Pythagorean Theorem:

"The sum of the square roots of two sides of an isosceles triangle is equal to the square root of the remaining side."

Name three things wrong with Homer's rendition.

65. Name three things that show that the following triangle could not exist.

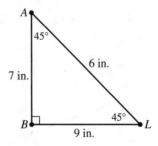

66. Name three things that show the following triangle could not exist.

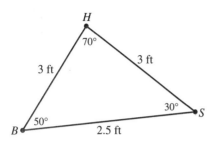

67. Explain why the following triangle could not exist.

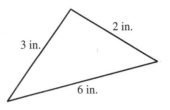

CONSTRUCTION

68. **Ruler and Protractor.** Use a ruler to create any isosceles triangle:

    1. Draw two line segments that have the same length and share an endpoint.

    2. Now connect the other two endpoints to create the third side of the triangle.

    In this section, we mentioned that in an isosceles triangle, angles opposite congruent sides have the same measure. Use a protractor to verify the statement for your isosceles triangle by measuring and labeling all three angles in your triangle.

69. **Ruler and Protractor.** Construct a triangle using the following steps.

    (1) Draw a line segment $\overline{SH}$ that is exactly 3 inches.

    (2) Draw a line segment $\overline{SB}$ so that $\overline{SB}$ is exactly 5 inches and so that the $m\angle HSB$ is 46°.

    (3) Draw line segment $\overline{HB}$ to complete the triangle.

    Use a ruler and protractor to answer the following questions.

    a. What is the length of △HSB's third side?

    b. What is the measure of the remaining two angles in △HSB?

    c. How is △HSB classified according to its sides?

    d. How is △HSB classified by its angles?

70. **Ruler and Compass.** Use a ruler and a compass (but not a protractor!) to create an equilateral triangle that has sides that measure three inches as outlined in the Construction Example in this section.

71. **Straightedge and Compass.** Follow these steps carefully.

    (1) Create any scalene triangle with sides anywhere between 5 and 10 centimeters.

    (2) Use the construction technique from Section 1 about dividing line segments to find the midpoint of each side of the triangle.

    (3) For each midpoint, draw a line segment from the midpoint to the opposite vertex (recall that these are called *medians*). Draw all three medians of the triangle.

    (4) If this is done correctly, all three medians will intersect in one point. Mark this point.

    The point where the medians of a triangle intersect is called the *centroid* of the triangle. You can perfectly balance any triangle on the tip of a pencil by placing the tip at the centroid.

| Geometry Project 3 | Proofs in Geometry

In the previous Geometry Project, we introduced the process of proving theorems in geometry. As new concepts and definitions are explored, more theorems can be proven. All the theorems in this project involve information about triangles.

As in the project before, we will show an example of how a two-column proof about triangles would be organized and how it helps in doing proofs for yourself.

Theorem: *The sum of the interior angles in a triangle is 180°.*

In the text we gave an illustration of how this works, but that illustration is not *proof.* Here is a two-column proof for the theorem.

Prove: $m\angle 1 + m\angle 2 + m\angle 3 = 180°$.

Set up: Draw a line parallel to the base of the triangle and label the resulting angles for reference.

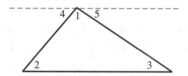

Statement	Justification
$m\angle 4 + m\angle 1 + m\angle 5 = 180°$	There are 180° in a straight angle (the addition property of angles).
$m\angle 4 = m\angle 2$ $m\angle 5 = m\angle 3$	Alternate interior angles theorem
$m\angle 2 + m\angle 1 + m\angle 3 = 180°$	The substitution property ∎

As before, in the exercises that follow you will help provide justification for a proof and then try to prove two theorems on your own.

Project Exercises | Writing Geometry Proofs

1. Consider the following figure.

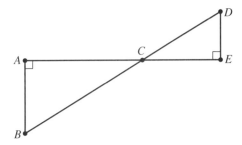

The following proof shows that $\angle CDE \cong \angle ABC$. The statements have been provided. For each statement, provide a justification. You may want to review the techniques from the project at the end of Section 2: More about Angles.

Statement	Justification
$m\angle BAC = m\angle DEC$	?
$m\angle ACB = m\angle DCE$	?
$m\angle ABC + m\angle BAC + m\angle ACB = 180°$ $m\angle CDE + m\angle DEC + m\angle ECD = 180°$	?
$m\angle ABC = 180 - m\angle BAC - m\angle ACB$ $m\angle CDE = 180 - m\angle DEC - m\angle ECD$	?
$m\angle ABC = 180 - m\angle DEC - m\angle ECD$	?
$\angle CDE \cong \angle ABC$	? ∎

2. **Theorem:** *The measure of an exterior angle of a triangle is equal to the sum of the two interior angles not corresponding to that exterior angle.* Use a two-column system to provide a proof of this theorem. It may help to begin by drawing a corresponding picture to illustrate.

3. **Theorem:** *In an isosceles triangle, the angle bisector of the noncongruent angle is perpendicular to the side it meets.* Use a two-column system to provide a proof of this theorem.

Strategy hint: Begin by drawing a corresponding figure. Which angles in the figure are automatically congruent by the definition of *angle bisector* and a *property of isosceles triangles*? Prove that the other two must be congruent. You might use the definition of *supplementary angles*.

More about Triangles—Congruence and Similarity

The illustration depicts several sailors stranded on a deserted island. They are using shadows to create triangles that will tell them their latitude and longitude. This method involves techniques discussed in this section (properties of similar triangles).

- Congruent Triangles
- Congruence Theorems for Triangles
- Similar Triangles
- Special Right Triangles

If △ABC is congruent to △DEF, then corresponding angles are congruent:

∠A ≅ ∠D
∠B ≅ ∠E
∠C ≅ ∠F

And corresponding sides are congruent:

$\overline{AB} \cong \overline{DE}$
$\overline{AC} \cong \overline{DF}$
$\overline{BC} \cong \overline{EF}$

At the beginning of Section 3: Triangles, we began by making the statement that many distances that would normally be difficult (or even impossible) to find can be found fairly easily using triangles. In this section we will explain exactly how. And, as promised, for the project at the end of this section you will have the opportunity to demonstrate the power of triangles yourself by using them to find the height of any building you choose. The techniques involved with such a project are based on a concept you would study in any pre-algebra course: ratios and proportions. The concepts arise by comparing two triangles. We begin by looking at two triangles that have the same shape and size. Then we will look at two triangles that have the same shape, but not necessarily the same size.

■ Congruent Triangles

Recall that any two geometric figures are called **congruent** if they are the exact same size and shape. For two triangles, this means that corresponding line segments have the same lengths and corresponding angles have the same degree measure. Here is a figure to illustrate.

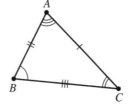 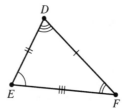

For notation, we write △ABC ≅ △DEF. Note that corresponding vertices are listed in the same order.

How is it possible to determine whether two triangles are congruent? We could measure all of the sides and all of the angles and compare them, but it is also possible to show that two triangles are congruent with less work and/or less information. We now present a set of theorems that can be used to show two triangles are congruent. We will precede each theorem with an explanation of why it is true.

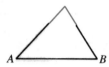

Congruence Theorems for Triangles

If the length of line segment $\overline{AB}$ and the measures of $\angle A$ and $\angle B$ are given, there is only one possible way to form a triangle with them. So if two triangles share these measurements, they must be congruent.

> **Angle-Side-Angle (ASA) Theorem:** *If two triangles have a corresponding side with equal measure and the two angles that share that side are congruent, then the two triangles must be congruent.*
>
>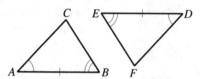

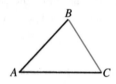

Now suppose that the measure of one angle and the lengths of the adjacent sides are given. Again there is only one way to complete a triangle.

> **Side-Angle-Side (SAS) Theorem:** *If two triangles have two sides congruent and the included angles are congruent, then the two triangles must be congruent.*
>
>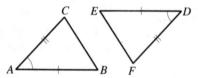

> ! This last principle explains why triangles are often used for bracing in construction. If the sides of a triangle are firmly attached at the vertices, there is only one possible shape for the triangle; the angles cannot be changed. This is not the case with four-sided objects; a four-sided object with fixed sides can have many different angle measures. So the design below is easy to deform:
>
>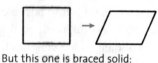
>
> But this one is braced solid:
>
>

Finally, if three line segments are given, there is at most one way to form a triangle with them.

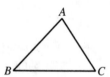

> **Side-Side-Side (SSS) Theorem:** *If two triangles have all three corresponding sides congruent, then the two triangles are congruent triangles.*
>
>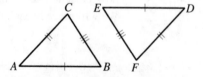

Notice from these theorems that the positioning of the angle and sides is important. That is, there is a difference between SAS: Side-Angle-Side and SSA: Side-Side-Angle. In SAS, the given sides make up the angle. In SSA, the two given sides do not make up the angle. SSA is not a sufficient condition for congruence.

| EXAMPLE 4–1 | Determining If Triangles Are Congruent |

The following figure shows four pairs of triangles. In each case, tell whether the triangles are necessarily congruent and why.

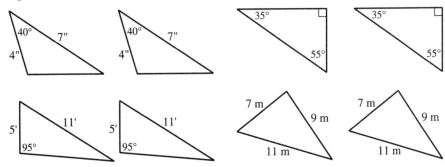

◆ **SOLUTION**

Here is the same figure with the results given below each pair of triangles.

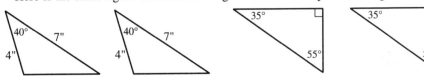

Yes, these are congruent by the SAS theorem.

No, these triangles share AAA. This is not sufficient for congruence.

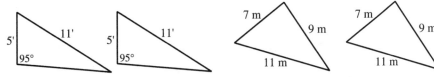

No, these triangles share SSA. This is not sufficient for congruence.

Yes, these are congruent by the SSS theorem.   ◆

Understanding congruence is an important start, but the most powerful applications of triangles come from exploring the situation where two triangles have the same shape, but not necessarily the same size.

■ Similar Triangles

Two geometric figures are **similar** if they have the same shape, but not necessarily the same size. The term *similar* can be applied to any two objects that have the same shape. For example, we use this concept when we make scale models, or use blueprints.

But in this section, we will explore the results of similarity specific to triangles, using the following two facts:

1. Corresponding sides of similar shapes are proportional.

2. Corresponding angles of similar shapes are congruent.

So if two triangles are similar, the three corresponding pairs of angles will have the same measure. The corresponding sides will probably not be equal, but if we divide the length of one side by the length of the corresponding side, we should get the same ratio for all three pairs. Here is an illustration of what we mean.

If $\triangle ABC$ is similar to $\triangle DEF$, then corresponding angles are congruent:

$\angle A \cong \angle D$
$\angle B \cong \angle E$
$\angle C \cong \angle F$

And corresponding sides are proportional:

$$\frac{AB}{DE} = \frac{BC}{EF} = \frac{AC}{DF}$$

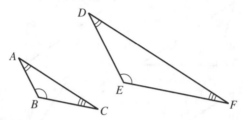

For notation, we write $\triangle ABC \sim \triangle DEF$. Suppose we take a diagram of $\triangle ABC$ and enlarge it to 200% on a copy machine, and the result is $\triangle DEF$. Then the corresponding angles are congruent, but the corresponding side lengths differ by a factor of 2. In general, if two triangles are similar and one side is k times the corresponding side of the other triangle, then the other two sides have been multiplied by k as well. The quantity k is often called the **scaling factor.**

EXAMPLE 4-2   Using Proportions to Find Missing Values

If the triangles below are known to be similar, find the values of x and y.

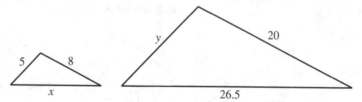

◆ **SOLUTION**

The values of x and y can be found from the proportion of the corresponding sides:

$$\frac{5}{y} = \frac{x}{26.5} = \frac{8}{20}$$

We use one proportion at a time:

$$\frac{x}{26.5} = \frac{8}{20}, \text{ so } x = 10.6 \qquad \frac{5}{y} = \frac{8}{20}, \text{ so } y = 12.5$$ ◆

How can we tell whether two triangles are similar? There are two theorems regarding similarity that we can use. They look similar to the theorems presented regarding congruency, so be careful not to confuse them.

AA Similarity Theorem: *If two triangles have two corresponding angles that are congruent, then the two triangles are similar.*

Note that if two of the angles are congruent, then the third angles are automatically congruent. Why?

SSS Similarity Theorem: *If two triangles have all three corresponding sides that are proportional, then the two triangles are similar.*

EXAMPLE 4–3   Finding Missing Values

Find x and y in the following figure.

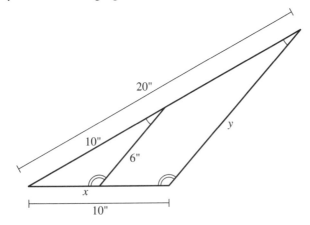

◆ SOLUTION

The figure contains two triangles that share one angle and have the other two marked as congruent. So the triangles are similar. Here is the ratio formed by the corresponding sides:

$$\frac{10}{20} = \frac{x}{10} = \frac{6}{y}$$

As before, we solve each proportion separately:

$$\frac{10}{20} = \frac{x}{10}, \text{ so } x = 5" \qquad \frac{10}{20} = \frac{6}{y}, \text{ so } y = 12" \qquad ◆$$

If a line segment parallel to one side of a triangle connects the other two sides, then a smaller triangle is formed within the larger one. The small and the large triangles are similar because the parallel lines form congruent corresponding angles. This means the triangles are similar by the AA similarity theorem.

If we look at triangles formed by shadows of objects at the same time of day, these triangles are also similar because the angle of elevation of the sun is the same in both triangles. This idea leads to a way to find the height of extremely tall objects relatively easily.

EXAMPLE 4-4   Finding the Height of a Flagpole

Wayne and Holly have a bet. Holly claims she can find the height of a tall flagpole using only a yardstick and Wayne doesn't believe she can. Holly has Wayne stand perfectly still and measures the length of his shadow. It is 4 feet 6 inches, or 4.5 feet. She then finds that Wayne is exactly 6 feet tall. Finally, she uses the yardstick to find that the flagpole casts a shadow 24 feet long. How tall is the flagpole?

◆ SOLUTION

The scenario in Holly's mind is illustrated below (not drawn to scale).

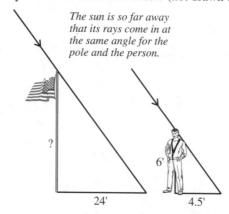

The sun is so far away that its rays come in at the same angle for the pole and the person.

Holly has created a situation that involves two similar triangles. She can now establish a proportion between the corresponding sides:

$$\frac{4.5}{24} = \frac{6}{x}, \text{ so } x = 32$$

The height of the flagpole is 32 feet (and Wayne loses the bet).   ◆

There are other situations in which we rely on properties of angles to conclude that two triangles are similar. This may involve reflections, or vertical angles, or properties of parallel lines being cut by a transversal (as described in Section 2: More about Angles).

EXAMPLE 4-5   Finding Missing Values

Find the measure of each missing side in the following figure. Round your results to the nearest tenth, if necessary.

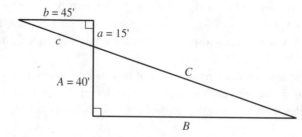

◆ **SOLUTION**

Both of the triangles have a right angle. In addition, two of the angles are vertical angles, which must be congruent. So the two triangles are similar by the AA similarity theorem. The orientation of the triangles provides a challenge. To help you see the sides correspond, we have labeled the sides of each triangle with upper- and lowercase letters:

$$a = 15 \rightarrow A = 40$$
$$b = 45 \rightarrow B = \;?$$
$$c = \;? \rightarrow C = \;?$$

We can set up a proportion to find B:

$$\frac{15}{40} = \frac{45}{B}, \text{ so } B = 120$$

Notice that neither c nor C are given, but that both are the hypotenuse of a right triangle. So we will use the Pythagorean Theorem to find one, and then a proportion to find the next. Let's use the Pythagorean Theorem to find c:

$$a^2 + b^2 = c^2$$
$$15^2 + 45^2 = c^2$$
$$c^2 = 2250$$
$$c = \sqrt{2250} \approx 47.4$$

Now that we have c, we can use it in a proportion to find C:

$$\frac{15}{40} = \frac{47.4}{C}, \text{ so } C = 126.4$$

Note that we could have also found C using the Pythagorean Theorem. Our final result is $B = 120$ ft, $c = 47.4$ ft, and $C = 126.4$ ft. ◆

■ Special Right Triangles

As we have seen before, the measure of the angles of a triangle and the lengths of the sides are related. For example, equilateral triangles have all angles congruent, and isosceles triangles have two congruent sides and angles. We have also seen that right triangles have an important relationship between the lengths of their sides; that is, the Pythagorean Theorem. We can combine the Pythagorean Theorem with the properties of isosceles and equilateral triangles and the concept of similarity to take these types of relationships further for certain special triangles.

Isosceles Right Triangle: 45°-45°-90°

Recall that an isosceles triangle has two congruent sides. In a right triangle, these must be the legs. Then the two acute angles must be congruent. Therefore, each of them must be a 45° angle.

In this case $a^2 + b^2 = c^2$ can also be written as $a^2 + a^2 = c^2$ because $a = b$. Then $2a^2 = c^2$, or $c = \sqrt{2a^2} = \sqrt{2}a$. This leads to the folowing result.

In any 45-45-90 triangle, the legs are congruent, and the hypotenuse is equal to $\sqrt{2}$ times the length of the legs.

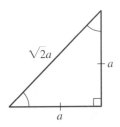

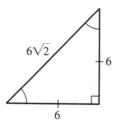

This is true for any isosceles right triangle because all isosceles right triangles have the same angles, so they are similar and their sides are proportional. As mentioned before, the result we just achieved involved using properties of isosceles triangles, the Pythagorean Theorem, and properties of similar triangles.

> **EXAMPLE 4-6**   Using Properties of a 45-45-90 Triangle

In the figure below, find x and y.

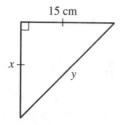

◆ **SOLUTION**

We begin by noting that the triangle is an isosceles right triangle. Therefore, it classifies as a 45-45-90 triangle. This makes finding the missing sides easy. The second leg is x, and it is congruent to the given leg, so $x = 15$ cm.

The hypotenuse is labeled y, so $y = \sqrt{2} \cdot 15 = 15\sqrt{2} \approx 21.2$ cm.   ◆

30°-60°-90° Triangle

Another important triangle has angle measures equal to 30°, 60°, and 90°. If we draw two of these back to back, the result looks like this:

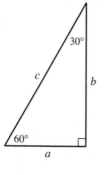

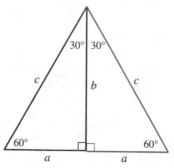

The outer border of the figure forms an equilateral triangle because all of the angles measure 60°. Then all of the sides are congruent so $c = a + a = 2a$, or just $c = 2a$. So the hypotenuse of a 30-60-90 is twice as long as the shorter leg.

In order to calculate the length of side b, we use the Pythagorean Theorem as follows:

$$a^2 + b^2 = c^2$$
$$a^2 + b^2 = (2a)^2 \text{ since } c = 2a$$
$$b^2 = 4a^2 - a^2 = 3a^2$$
$$b = \sqrt{3}a$$

So the longer leg in a 30-60-90 is equal to the shorter times $\sqrt{3}$.

Again these relationships are true for all 30-60-90 triangles because they are all similar and therefore their sides are proportional. The triangle in the margin illustrates these relationships.

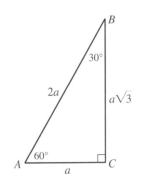

In any 30-60-90 triangle, the leg opposite the 60° angle is $\sqrt{3}$ times the length of the leg opposite the 30° angle, and the hypotenuse is equal to twice the length of the leg opposite the 30° angle.

EXAMPLE 4-7   Using Properties of a 30-60-90 Triangle

Given $\triangle ABC$ below is a 30-60-90 triangle, and $AB = 20$, find the lengths of $\overline{AC}$ and $\overline{AB}$.

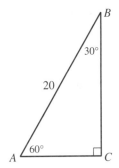

◆ **SOLUTION**

By using the triangle as described above, we can find that $AC = 10$ since it is half the length of side $\overline{AB}$, and $BC = 10\sqrt{3} \approx 17.32$.   ◆

Applications of Special Triangles

A surveyor has been asked to find the height of a large tree. To do this, he or she need only use a part of what you learned in this section and a little innovation. To find the height, the surveyor uses a device called a *theodolite* to find the place where the angle of elevation from himself or herself to the tree is exactly 45°. Because the tree is standing perpendicular to the ground, the surveyor now knows he or she has created a 45-45-90 triangle. In this section, we learned that if you know one side in a 45-45-90 triangle you can find the other two. The length of the side opposite the 45° angle at the theodolite is 32 feet, which happens to be the missing length. Now all the surveyor has to do to get the total height of the tree is to add 4 feet (labeled on the lower left side of the trunk) to compensate for the height of the theodolite. So the total height of the tree is 36 feet.

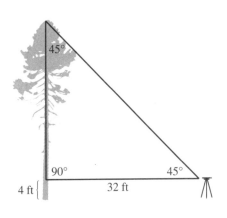

EXAMPLE 4-8   Supporting a Tower

A radio tower is 120 feet tall. To help support the tower a wire will be attached to the top of the tower and secured to the ground. The wire is to be positioned so that it inclines 60° off the ground. How long will the wire be? Give an exact value and approximate to the nearest tenth.

◆ **SOLUTION**

Compare the figure below to the 30-60-90 triangle as shown previously.

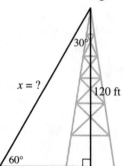

From the figure, we know that when a ratio (fraction) is formed out of the leg opposite the 60° angle and the hypotenuse we must get $\frac{\sqrt{3}}{2}$. So in our figure when we divide those sides we have to get the same thing. That is,

$$\frac{120}{x} = \frac{\sqrt{3}}{2} \qquad \textit{Cross multiply}$$

$$\Rightarrow 240 = \sqrt{3}x \qquad \textit{Divide both sides by } \sqrt{3}$$

$$\Rightarrow x = \frac{240}{\sqrt{3}} = \frac{240}{\sqrt{3}} \cdot \frac{\sqrt{3}}{\sqrt{3}} = \frac{240\sqrt{3}}{3} = 80\sqrt{3} \text{ ft}$$

Note that we rationalized the denominator in this last step. So the wire will be $80\sqrt{3}$ ft ≈ 138.6 ft.

◆

Section 4 | Exercises

For exercises 1–4, use the terms from the Vocabulary Checklist to fill in the blanks.

✔ **VOCABULARY CHECKLIST:**

vcongruent triangles

ASA for Congruence

SAS for Congruence

SSS for Congruence

similar triangles

scaling factor

AAA for similarity

SSS for similarity

1. A(n) _____ is a single number that gives the relationship between sides of two similar triangles.

2. The two theorems given that involve similarity can be described using the letters _____ and _____.

3. Two triangles that have the same shape are called _____.

4. _____ have exactly the same side and angle measures.

For exercises 5–8, answer TRUE or FALSE.

T | F   5. Anytime two triangles have two congruent sides and one congruent angle they will be congruent.

T | F   6. Anytime two triangles have one congruent side and two congruent angles they will be congruent.

T | F   7. A pair of similar triangles can always be considered congruent.

T | F   8. A pair of congruent triangles can always be considered similar.

9. Explain the difference in the two following notations:
$\triangle ABC \cong \triangle DEF$ and $\triangle ABC \sim \triangle DEF$

10. Given the figure below, fill in the blanks in the statements that follow.

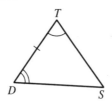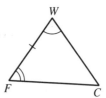

$\triangle SDT \cong$ _____

$\overline{FW} \cong$ _____

$\overline{TS} \cong$ _____

$\angle D \cong$ _____

$\angle C \cong$ _____

11. Given the figure below, fill in the blanks in the statements that follow.

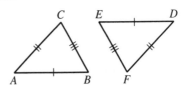

$\triangle BAC \cong$ _____

$\overline{AB} \cong$ _____

$\overline{FD} \cong$ _____

$\angle B \cong$ _____

$\angle D \cong$ _____

12. The figure below shows similar triangles. Fill in the blanks in the statements that follow.

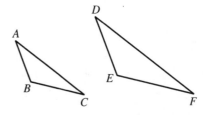

$\triangle ABC \sim$ _____

$\angle BAC \cong$ _____

$\angle EFD \cong$ _____

$$\frac{AB}{?} = \frac{?}{DF} = \frac{BC}{?}$$

13. The figure below shows similar triangles. Fill in the blanks in the statements that follow.

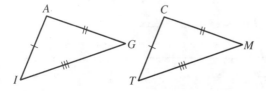

$\triangle NDY \sim$ _____

$\angle LKN \cong$ _____

$\angle NDY \cong$ _____

$$\frac{LK}{?} = \frac{?}{ND} = \frac{KN}{?}$$

14. For the following figure, name the congruent triangles and write six congruence statements that pertain to the triangles.

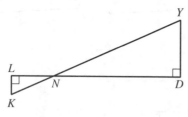

15. For the following figure, name the congruent triangles and write six congruence statements that pertain to the triangles.

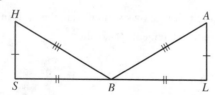

16. Assuming the two triangles shown are similar, write three similarity statements for the angles and a three-part pro-portion for the sides.

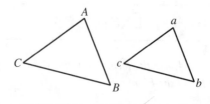

17. For the following figure, state which triangles are similar. Write three similarity statements for the angles and a three-part proportion for the sides. Assume that $\overline{JW}$ is parallel to $\overline{EM}$.

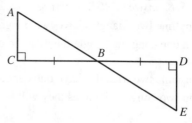

18. Tell why $\triangle ABC$ and $\triangle ADC$ are congruent.

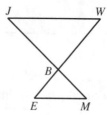

19. Tell why $\triangle ACB$ and $\triangle EDB$ are congruent.

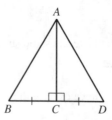

20. Why are the following triangles not congruent even though they have two congruent sides and one congruent angle?

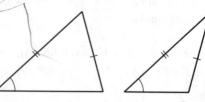

21. Explain why $\triangle ABC$ and $\triangle DEC$ are similar and write an appropriate similarity statement.

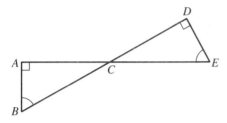

22. Explain why $\triangle ABC$ and $\triangle DCE$ are similar and write an appropriate similarity statement. Assume $\overleftrightarrow{AB}$ and $\overleftrightarrow{DE}$ are parallel.

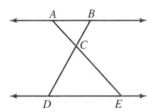

23. Explain why $\triangle ABD$ and $\triangle ACE$ are similar and write an appropriate similarity statement. Assume $\overleftrightarrow{BD}$ and $\overleftrightarrow{CE}$ are parallel.

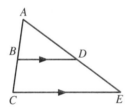

24. Explain why these triangles are similar.

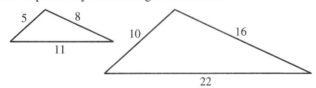

25. In the following figure, find b and f.

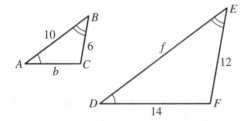

26. In the figure below, find x and y. Assume the appropriate line segments are parallel.

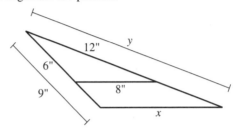

27. If $\overline{BE}$ is parallel to $\overline{CD}$, find x and y.

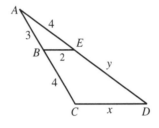

28. Find b, c, and B.

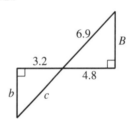

29. **Finding the Height of a Tree.** A forest ranger is trying to find the height of a tree. He is $5\frac{1}{2}$ feet tall. He notes that the length of his shadow is 11 feet and that the length of the tree's shadow is 23 feet. How tall is the tree?

30. **Very Easily Done.** As a mathematical experiment, a student goes outside on a sunny day, takes a yardstick (which is 3 feet long), stands it straight up, and measures its shadow, finding that the shadow is only 8 inches long. She then goes around and measures the shadows of different objects. Her results are given below. Finish the table.

Object	Shadow Length	Height of Object
Stop sign	1'6"	
Building	5'4"	
Tree	6'	
Pepsi bottle	1"	

31. **Using Each Other to Find Heights.** On a sunny day, Michelle and Nancy noticed that their shadows were different lengths. Nancy measured Michelle's shadow and found that it was 96 inches long. Michelle then measured Nancy's shadow and found that it was 102 inches long. Who do you think is taller, Nancy or Michelle? Why? If Michelle is 5 feet 4 inches tall, how tall is Nancy? If Nancy is 5 feet 4 inches tall, how tall is Michelle?

32. If $\overline{AB}$ is parallel to $\overline{DE}$, $m\angle A = 77°$, and $m\angle M = 57°$, find the following: $m\angle B$, $m\angle D$, $m\angle E$, a, and b.

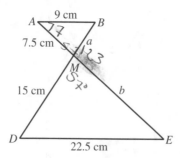

33. **Finding a Distance.** A lamp mounted on a building 12 feet above the ground shines over a 9-foot pole. The shadow of the pole is 15 feet long. How far is the pole from the building? Assume the lamp is flush with the building.

34. **Finding the Height of a Building.** Holly and Wayne are at it again. This time Holly claims she can find the height of Wayne's house using only a measuring tape and a mirror. She places the mirror on the ground somewhere in front of the building and backs away from it until she can see the top of the building in the mirror. This results in the following scenario (not drawn to scale). Find the height of the house.

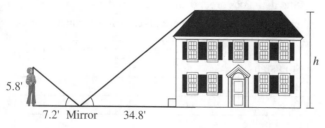

35. Two right triangles are similar. The short leg of the small triangle is 6 inches long. The short leg of the larger triangle is 18 inches long. The longer leg of the small triangle is 10 inches long. What is the scaling factor?

36. Give the length of the other two sides.

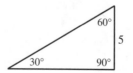

37. Give the length of the other two sides.

38. Give the length of the other two sides.

39. Give the length of the other two sides.

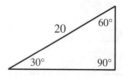

40. Give the length of the other two sides.

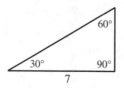

For exercises 41–44, sketch the triangle described and label each side and angle. When giving the side lengths, give an exact value, then round to the nearest tenth.

41. A 30-60-90 where the side opposite the 30° angle is 5 cm.

42. A 45-45-90 where the side opposite a 45° angle is 3 in.

43. A 45-45-90 where the side opposite a 90° angle is 10 in.

44. A 30-60-90 where the side opposite the 60° angle is 5 ft.

45. **Surveying.** A surveyor trying to find the height of a building walks away from it until his line of sight to the top of the building makes a 60° angle with the ground. He marks this point and finds it is 54 feet from the building. How tall is the building?

46. **The Height of a Weather Balloon.** A weather balloon is tied to the end of 76 feet of string. If the string makes a 45° angle with the ground, how high off the ground is the weather balloon?

CONSTRUCTION

47. **Ruler and Protractor.** Use a protractor and straightedge to draw a triangle that has angles measuring exactly 30, 60, and 90 degrees. Now use a ruler and a calculator to verify the following statement:

    The length of the longer leg of a 30°-60°-90° triangle is the length of the shorter leg times $\sqrt{3}$. The hypotenuse is twice as long as the shorter leg.

48. Repeat exercise 47 for a 45-45-90 triangle using the following statement:

    In an isosceles right triangle, the length of the hypotenuse is the length of a leg times $\sqrt{2}$.

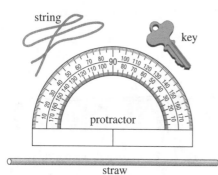

string

key

protractor

straw

Geometry Project 4   Finding the Height of a Building

At the beginning of Section 3, you were told that triangles are so practical that by the end of Section 4 you could use household items to find the height of a building. For this project you will do just that. You will use the following items to make a *sextant*. This is an object that will measure an angle of elevation to the top of a building.

Items You Will Need

- a protractor
- a straw
- tape
- two feet of string
- a small weight such as a washer or key
- a measuring tape or a yardstick
- a building (for which you will find the height)
- a partner (taking the measurements will be easier if you work with someone)

How to Make a Sextant

1. Tape the straw to the base of the protractor to use as a sight. Be careful, and make sure the straw is indeed parallel to the base when you tape it down or your sight will be crooked.

2. Tie the washer to one end of the string.

3. Tie the second end to the protractor at the center of the base. The string will serve as the measure mark to make it easy to read an angle of elevation.

How to Find the Height of a Building

1. Find the distance from the building for which the angle of elevation is exactly 60°. This will probably mean starting at the base of the building and backing up, periodically stopping to find the angle of elevation using your homemade sextant. (Keep in mind that 90° on the protractor corresponds to a 0° angle of elevation. So 30° on the protractor corresponds to a 60° angle of elevation.)

2. Once you have found that point, measure the distance from that point to the base of the building.

3. You have created a 30-60-90 triangle. Fill in the known measurement and use properties of special triangles to find the missing height. If necessary, round your answers to the nearest tenth.

Write a report summarizing your method and result. Discuss any problems that came up in the process. Also address what problem(s) you foresee in trying this technique in other situations.

Quadrilaterals

Quadrilaterals involve four vertices, four line segments, and four angles. A quadrilateral can be formed from any four points so long as no three of them lie on the same straight line.

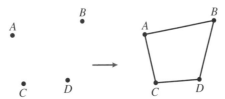

Note that while the figure below has four labeled points and four line segments, it is not a quadrilateral. The sides of a quadrilateral cannot cross.

Quadrilaterals are obviously a part of everyday life (just look around). But their practical application goes far beyond the obvious square and rectangle constructions you see everywhere you look. Here is an example that would be explored in more detail in a course on trigonometry.

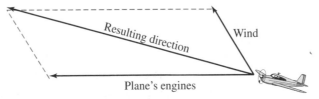

Quadrilaterals can be used to determine the result from wind pushing on an airplane. The sides of the **parallelogram** represent the two forces pushing the plane: its own engines and the wind. The **diagonal** of the parallelogram represents the resulting direction. We will discuss these terms more in a bit.

■ Classifying Quadrilaterals

As with triangles, we begin by classifying quadrilaterals according to their angles and sides. We also use the accompanying table to explain the terminology involved with these shapes and some of their properties.

SPECIAL TYPES OF QUADRILATERALS

CLASSIFICATION	DESCRIPTION/PROPERTIES/TERMINOLOGY	EXAMPLE
Parallelogram	Opposite sides of the quadrilateral are parallel. Properties: • Opposite sides are congruent. • Opposite angles are congruent. • Angles that share a side are supplementary. • The parallelogram's **perpendicular height** is indicated with a dotted line.	
Rectangle	All four angles in the quadrilateral are right angles. Properties: • Opposite sides are parallel. • Opposite sides are congruent. • Automatically a parallelogram.	
Rhombus	All four sides are the same length. A rhombus is automatically a parallelogram.	
Square	All four sides are the same length and all four angles are right angles. A square qualifies as a rectangle, a parallelogram, and a rhombus.	
Trapezoid	A quadrilateral with two parallel sides. Terminology: • The parallel sides are referred to as the **bases.** • The other two sides are referred to as the **legs.** • The trapezoid's **perpendicular height** is indicated with a dotted line.	
Kite	A quadrilateral with two pairs of adjacent, congruent sides. Opposite angles created by noncongruent sides are congruent.	

Notice in the photo of the ancient Mayan temple at the beginning of this section that the sides are made of several trapezoids (eight total) and not just one big triangle. Let's look at a few examples that involve different ideas all related to quadrilaterals. The first is an exercise in understanding the definitions given for each quadrilateral.

EXAMPLE 5-1   Which Is Which?

In each of the following statements, fill in the blank with either "sometimes" or "always."

    a. A square is _____ a rectangle.

    b. A parallelogram is _____ a rectangle.

    c. A rhombus is _____ a parallelogram.

    d. A kite is _____ a rhombus.

◆ **SOLUTION**

    a. To be a rectangle, a quadrilateral must have four right angles. Since a square always has four right angles, a square is *always* a rectangle.

    b. A parallelogram has opposite sides congruent, but a rectangle has opposite sides congruent and four right angles. This could happen with a parallelogram, but not always. So a parallelogram is *sometimes* a rectangle.

    c. As mentioned in our table, a rhombus is *always* a parallelogram.

    d. A kite has two sets of adjacent, congruent sides. There is nothing that says that all four can't be congruent, so a kite is *sometimes* a rhombus.   ◆

Now let's look at an example that pulls from the information learned in Section 2 about parallel lines.

EXAMPLE 5-2   Finding Missing Angles

Find the measures of $\angle KMS$ and $\angle KBS$ in the following trapezoid.

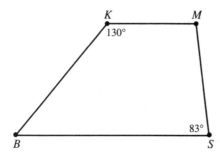

◆ **SOLUTION**

    You may want to review the topic "Parallel Lines Cut by a Transversal" in Section 2 before you look at this solution. There is no property of trapezoids that says the angles have to be congruent. However, the small and large bases are parallel. So line segments $\overline{MS}$ and $\overline{KB}$ are transversals to line segments $\overline{KM}$ and $\overline{BS}$. This means that the angles $\angle BKM$ and $\angle KBS$ are same-side interior angles and so are angles $\angle KMS$ and $\angle BSM$. As we have seen, same-side interior angles are supplementary, so we can create and solve the following equations:

$$m\angle BKM + m\angle KBS = 180° \qquad m\angle KMS + m\angle BSM = 180°$$
$$130° + m\angle KBS = 180° \qquad m\angle KMS + 83° = 180°$$
$$m\angle KBS = 50° \qquad m\angle KMS = 97° \qquad ◆$$

Finally, we look at an example that involves using a ruler to create a quadrilateral to specifications.

EXAMPLE 5-3   Using a Ruler to Create an Isosceles Trapezoid

Create a trapezoid with the following properties:
- The length of the longer base is 8 centimeters.
- The length of the shorter base is 4 centimeters.
- The perpendicular height of the trapezoid is only 1 centimeter.
- The two legs of the trapezoid are equal in length.

◆ **SOLUTION**

Since the figure is to be a trapezoid, the bases must be parallel. We begin by drawing the 8-centimeter lower base and the perpendicular height close to the middle as a reference. Remember, the perpendicular height is not actually part of the trapezoid, but it is an important measurement:

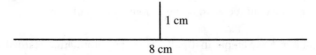

To make sure that the two legs are equal in length, it will be necessary to draw the upper base so that the difference between the upper base and the lower base is split between the ends of the longer base. A diagram should help. Look at the one below.

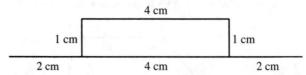

By setting the diagram up this way, we have ensured that when we draw the two legs of the trapezoid, we are actually drawing two congruent triangles. Since two sides are congruent, the third will be as well (this is easy to verify using the Pythagorean Theorem).

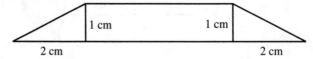

We could find the lengths of these two new sides very easily, but we weren't asked for their length; only to ensure that they are congruent. We finish by redrawing the trapezoid without all the reference lines used to create it.

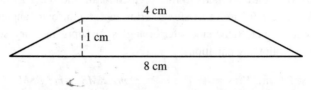

◆

A trapezoid with congruent legs is called an **isosceles trapezoid.** Make sure to reference this example when you work the corresponding exercise at the end of this section.

▪ The Angles of a Quadrilateral

The sum of the interior angles in a quadrilateral is always 360°. As with the triangle you are invited to test this notion using a paper quadrilateral. Cut out the corners and assemble them so they are adjacent. In every case this will lead to a total of 360°.

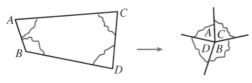

$$m\angle A + m\angle B + m\angle C + m\angle D = 360°$$

Another way you could show that the sum of the angles in a quadrilateral is 360° is to note that every quadrilateral can be divided into two triangles.

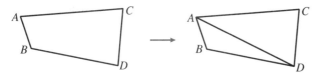

Since each one of the triangles must have a total of 180°, the quadrilateral must have a total of $2 \cdot 180° = 360°$.

EXAMPLE 5-4   Using Algebra with the Angles in a Quadrilateral

Given the following figure, find the value of x.

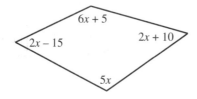

◆ SOLUTION

As we have seen with a triangle, having these expressions for the angle measures leads to the equation

$$(6x + 5) + (2x + 10) + (5x) + (2x - 15) = 360$$
$$15x = 360$$
$$x = 24$$

Notice that we were not asked to actually find the measure of each angle, so we are finished.   ◆

▪ Diagonals

A line segment that connects two vertices of a quadrilateral but does not make up a side is called a **diagonal.** The example given in the introduction to this section shows a practical use of diagonals. Diagonals are also used in construction to determine if a

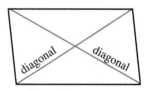

quadrilateral is really a rectangle or just looks like it is. There are several theorems that deal directly or indirectly with the diagonals of a quadrilateral. Here are some of the more important ones:

- The diagonals of a parallelogram bisect each other.
- The diagonals of a rhombus are perpendicular and bisect the angles at each vertex.
- The diagonals of a rectangle are congruent.

EXAMPLE 5–5　Marking a Parallelogram

The figure below is a parallelogram. Draw its diagonals and use tick marks to indicate all sets of congruent sides.

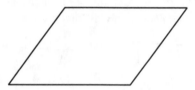

◆ SOLUTION

Here is the figure with the diagonals drawn and the congruent line segments marked.

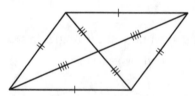

We use the fact that opposite sides of a parallelogram are congruent and the theorem above that says the diagonals bisect each other. Notice that this gives four pairs of congruent segments.　　　　　　　　　　　　　　　　　　　　　◆

EXAMPLE 5–6　OOPS!

A carpenter has framed a room that is to be 15 feet long and 9 feet wide. He remeasures all four walls and also measures the distance between opposite corners, with the result shown in the figure. What does this information tell the carpenter?

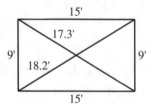

◆ SOLUTION

The goal in building a room like this is that the corners be right angles and therefore that the room is a true rectangle. Unfortunately, the diagonals of the quadrilateral did not come out to be equal. As the theorem mentions above, the diagonals of a rectangle would have been equal. So the carpenter has built a room that is not rectangular!　◆

Section 5 | Exercises

For exercises 1–4, use the terms from the **Vocabulary Checklist** *to fill in the blanks.*

✔

VOCABULARY CHECKLIST:

quadrilateral	square
rectangle	trapezoid
parallelogram	kite
rhombus	diagonal

1. Connecting two opposite vertices of a quadrilateral creates a _____.

2. The two special quadrilaterals that always involve all right angles are the _____ and the _____.

3. The only special quadrilateral that does not necessarily involve any parallel lines is the _____.

4. The two special quadrilaterals that always involve four congruent sides are the _____ and the _____.

For exercises 5–8, answer TRUE or FALSE.

T | F̲ 5. A rhombus will always have four congruent angles.

T | F̲ 6. The diagonals of a rectangle are always perpendicular.

T | F̲ 7. The sum of the angles in a quadrilateral is always 180°.

T̲ | F 8. Some quadrilaterals fall into more than one special category (such as square and rhombus).

9. Sketch an example of a parallelogram.

10. Sketch an example of a rhombus.

11. Sketch an example of a kite.

12. Sketch an example of a square.

13. Sketch an example of a quadrilateral that does not classify as any of the special shapes discussed in this section.

14. Sketch an example of a rectangle.

15. Sketch an example of a trapezoid.

16. Sketch an example of an isosceles trapezoid.

In exercises 17–21, what type(s) of quadrilateral(s) is (are) implied by each figure?

17.
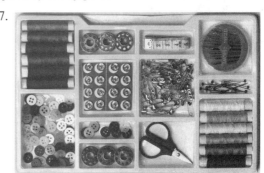

18.

19.

20.

21.

In exercises 22–28, fill in the blank with either "sometimes"
or "always."

22. A rectangle is _____ *a* a parallelogram.

23. A rhombus is _____ *s* a square.

24. A trapezoid is _____ *n* a parallelogram.

25. A rectangle is _____ *s* a square.

26. A square is _____ *a* a rhombus.

27. A kite is _____ *n* a parallelogram.

28. A quadrilateral is _____ *s* a rectangle.

For exercises 29–34, find the type of quadrilateral that could
fit the given description. There may be more than one for each
description.

29. Has two pairs of congruent sides.

30. All four angles are right angles.

31. At least two sides are congruent.

32. All four sides are congruent.

33. Has two pairs of congruent sides but no two sides are
     parallel.

34. Only one pair of sides is congruent.

35. Find the two missing angles and sides in the following
     kite.

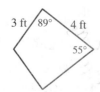

36. Find the two missing angles and sides in the following
     kite.

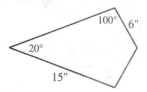

37. Find the two missing angles and sides in the following
     parallelogram.

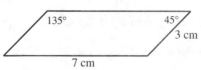

38. Find the two missing angles and sides in the following parallelogram.

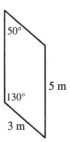

39. Label each angle and side of the rectangle, given that the longer side is three times the shorter side.

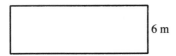

6 m

40. Label each angle and side of the rectangle, given that the shorter side is 5 less than half the longer side.

24 cm

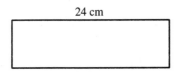

41. Find the two missing angles in the following trapezoid.

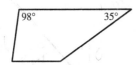

42. Find the two missing angles in the following trapezoid.

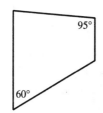

43. **Geometry and Algebra.** Use the following figure to find the value of x.

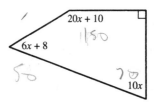

44. **Geometry and Algebra.** Use the following figure to find the value of x.

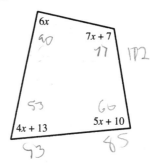

45. Solve the following problem, if possible:

   Two angles in a parallelogram are 75° and 105°. Find the measure of the other two angles.

46. Solve the following problem, if possible:

   Two angles in a trapezoid are 75° and 105°. Find the measure of the other two angles.

47. Sketch any square. Now draw its diagonals and use tick marks to label all congruent line segments in the figure.

48. Sketch a rhombus. Now draw its diagonals and use tick marks to label all congruent line segments in the figure.

Exercises 49–52 refer to the following rhombus.

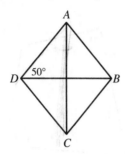

49. Find $m\angle DAC$.

50. Find $m\angle ADC$.

51. Find $m\angle DCB$.

52. Find $m\angle CAB$.

53. The smaller base of an isosceles trapezoid is 5 cm and the larger base is 11 cm. If the height of the trapezoid is 4 cm, what is the length of each one of the legs?

54. **The Median of a Trapezoid.** The **median** of a trapezoid is a line segment that bisects each of its legs. It can be shown that the median of a trapezoid is parallel to its bases. Use this information to sketch and label features for the median of the following trapezoid.

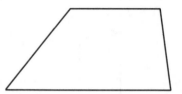

55. **The Median of a Trapezoid.** Repeat exercise 54 for the following trapezoid.

CONSTRUCTION

56. **Ruler and Protractor.** Create an exact square with side length 4 inches.

57. **Ruler and Protractor.** Create a parallelogram that has an angle of 105° made by sides of length 2 inches and 3 inches.

58. **Ruler.** Create a trapezoid with the following features:
    a. The longer base is 6 inches.
    b. The shorter base is 3 inches.
    c. The perpendicular height is 4 inches.
    d. The trapezoid is an isosceles trapezoid.

59. **Straightedge and Compass.** Construct a perfect rectangle without the use of a protractor.
    Hint: Use the construction technique from Section 1 for constructing a perpendicular bisector to create the right angles. It doesn't matter what the two side lengths are.

60. **Straightedge and Compass.** Construct a perfect rhombus without the use of a protractor.
    Hint: Start by drawing a line segment. Make the second segment of the angle by copying the first line segment as per the steps from the Construction Exercises in Section 1. Bisect this angle using the steps from Section 2. This line must also bisect the opposite angle, so the opposite vertex is on this line.

| **Geometry Project 5** | Quadrilaterals and Bisectors |

This project ties in several different skills and ideas:

- working with a ruler and protractor
- construction techniques from Section 1 using only a straightedge and compass
- basic concepts from Section 4
- basic concepts from this section

1. On a blank piece of paper, use a ruler and protractor to copy the following quadrilateral using the actual measurements given (centimeters).

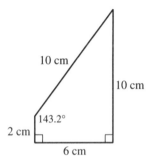

In Section 1, you were shown how to construct the midpoint of a line segment using a straightedge and compass. We briefly mentioned that this technique can also be used to draw a *perpendicular bisector* (a line or line segment that is perpendicular to a given segment and bisects the given line segment).

2. Use a straightedge, protractor, and the construction technique from Section 1 to draw the perpendicular bisectors of each of the sides of the quadrilateral. This will be most effective if you do it in a different color ink. Extend the perpendicular bisectors until each one intersects with another.

3. These four lines will create another quadrilateral; darken its sides so it can clearly be seen.

4. What appears to be the relation between this new quadrilateral and the original quadrilateral? How could this be verified using a ruler, a protractor, and perhaps a calculator? Make the measurements to verify the result.

5. Does your intuition suggest that this relationship will always be true or that this result was caused by the right angles? To test, repeat the process for another quadrilateral that you create on your own.

SECTION

6

Polygons

- Identifying and Classifying Polygons

- Theorems Concerning the Angles and Diagonals of a Polygon

Computer-generated images are popular in modern movies. The process of creating the images is based largely on manipulations of polygon surfaces. A cube is created with six square ~~sides~~ faces. A process called the *Catmull-Clark scheme* is used to repeatedly split each side into smaller polygons (usually quadrilaterals). The image can then be manipulated by moving a vertex of any polygon in the figure.

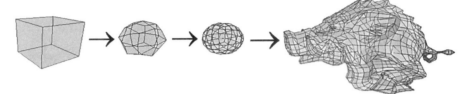

■ Identifying and Classifying Polygons

In the last two sections we discussed two examples of two-dimensional figures: triangles (three-sided) and quadrilaterals (four-sided). Triangles and quadrilaterals are two examples of polygons. In general, a **polygon** is any closed figure made of line segments that do not cross.

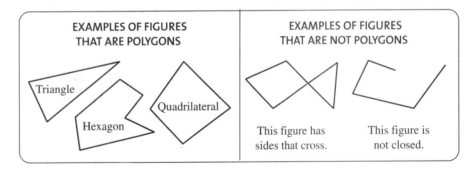

The word *polygon* comes from the Greek words for *many* (poly-) and *angle* (-gon). Although we usually name polygons for the number of sides, the number of sides and the number of angles is always the same. As before, the common points shared by line segments are called **vertices.**

In this section we will discuss properties that apply to all polygons. The properties will be similar in nature to the types of properties from Sections 3 and 4 on triangles and quadrilaterals.

NAMES OF POLYGONS	
NUMBER OF SIDES	NAME OF POLYGON
3	Triangle
4	Quadrilateral
5	Pentagon
6	Hexagon
7	Heptagon
8	Octagon
9	Nonagon
10	Decagon
n	n-gon

For polygons with more than four sides, we use the Greek names shown in the table in the margin. For polygons with more than 10 sides, we simply use the term **n-gon.** A **regular polygon** is one in which all sides and angles are equal in measure.

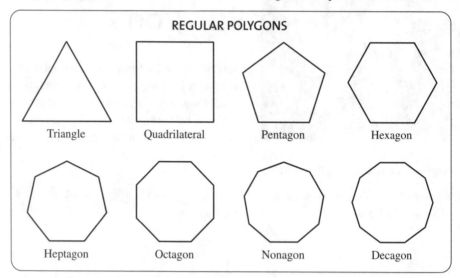

REGULAR POLYGONS

Triangle   Quadrilateral   Pentagon   Hexagon

Heptagon   Octagon   Nonagon   Decagon

As in Section 5, a **diagonal** of a polygon is a line segment joining two vertices of the polygon that are not joined by a side. That is, a line segment that joins two vertices and is not itself a side.

A polygon is **convex** if each of its diagonals lies completely inside the polygon. If the polygon is not convex, it is called **concave.** Some examples of convex and concave polygons are shown below.

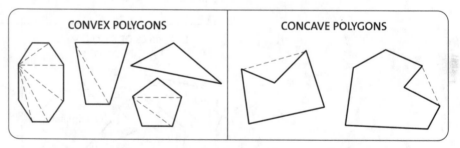

CONVEX POLYGONS   **CONCAVE POLYGONS**

Notice that for each of the concave polygons, there is a diagonal (shown as a dashed line) outside the polygon. In the convex polygons, all of the diagonals lie inside the polygon. For the remainder of this section, we will focus on convex polygons. Regular polygons, for example, are always convex polygons.

EXAMPLE 6–1   Sketching Polygons

Draw an example of a convex hexagon and a concave nonagon.

◆ **SOLUTION**

The figures can look like whatever we want, so long as they have the indicated features. A convex hexagon has six sides and the convex property.

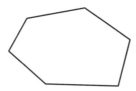

Note that all the diagonals of this polygon are contained in its interior. A concave nonagon has to have nine sides and the concave property.

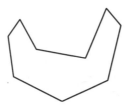

Note that some of the diagonals of this polygon would fall in the exterior of the object.   ◆

■ Theorems Concerning the Angles and Diagonals of a Polygon

Here we explore several theorems about convex polygons and provide a brief explanation about why the theorems are true. One of the "proofs" is even explored as a homework exercise. After we look at the theorems, we will explore their use with examples.

Diagonals of a Polygon *There are n − 3 diagonals coming from one vertex of an n-gon.*

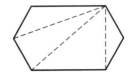

Starting at one vertex of an *n*-gon, we draw a diagonal to all but three of the *n* vertices: we do not draw a diagonal to the starting vertex, or to the two vertices directly adjacent to the starting vertex since they make sides. This is illustrated in the hexagon to the right.

Sum of the Angles in a Polygon *The sum of the angles of an n-gon is given by the formula S = 180° · (n − 2).*

Drawing all the diagonals from one vertex of a polygon divides the polygon into $n - 2$ triangles. You should draw a few examples to convince yourself of this fact. Since each triangle has 180° in it, the total number of degrees in the polygon is $180°(n - 2)$.

Angle Measures of a Regular Polygon *In a regular polygon, the measure of each angle can be found by the formula* $A = \dfrac{180°(n - 2)}{n}$.

The previous theorem says that the sum of all the angles is given by $180°(n - 2)$. In a regular *n*-gon there are *n* congruent angles. So the measure of each angle is $\dfrac{180°(n - 2)}{n}$.

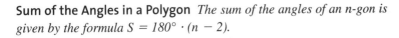

The Number of Sides in a Regular Polygon *Given that each angle in a regular polygon measures A degrees, the number of sides is given by the formula*
$$n = \frac{360°}{180° - A}.$$

This theorem is a direct result of the previous theorem. The algebraic manipulation is saved as a homework exercise.

EXAMPLE 6-2   Finding Properties of a Regular Polygon

For the polygon below, give the number of diagonals from one vertex and the measure of each interior angle.

◆ **SOLUTION**

This stop sign is a regular octagon, so $n = 8$.

There are $8 - 3 = 5$ diagonals from one vertex, and the measure of each interior angle is $\dfrac{180°(n - 2)}{n} = \dfrac{180° \cdot (8 - 2)}{8} = \dfrac{1080°}{8} = 135°$.   ◆

EXAMPLE 6-3   Finding the Number of Sides of a Polygon

A regular polygon has angles that measure $144°$. How many sides does the polygon have?

◆ **SOLUTION**

We were given the measure of each angle and asked to find the number of sides, so we use the formula $n = \dfrac{360}{180 - A}$ with $A = 144$.

$$n = \frac{360}{180 - 144} = \frac{360}{36} = 10$$

So the polygon has 10 sides; it is a regular polygon.   ◆

EXAMPLE 6-4   Finding the Total of the Angles in a Polygon

What would be the sum of all the angles in a regular 43-gon?

◆ **SOLUTION**

While the figure we are dealing with would be very difficult to draw, getting the answer to the question is straightforward. It's just an application of the formula $S = 180° \cdot (n - 2)$. Letting $n = 43$, we get $S = 180° \cdot 41 = 7380°$.   ◆

Finally, just as we considered exterior angles of a triangle, we can consider exterior angles for a polygon. In the case of a polygon, the exterior angles would be arranged as follows.

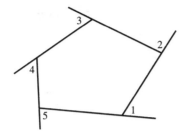

The Exterior Angles of a Polygon *The sum of the exterior angles of a polygon is always 360°, no matter how many sides the polygon has.*

Since exterior angles are supplementary to their corresponding interior angles, if the polygon has n sides, then there is a total of $180° \cdot n$ for the interior and exterior angles. As we have seen above, $180° \cdot (n - 2)$ of these are taken up by the interior angles. So that means there are $180° \cdot n - 180° \cdot (n - 2)$ left over for the exterior angles. But notice that $180° \cdot n - 180° \cdot (n - 2) = 180n - 180n + 360 = 360°$. This verifies the theorem.

This may seem counterintuitive, but compare the exterior angles in the following regular polygons.

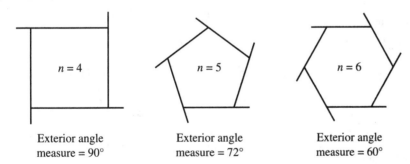

Notice that as the number of sides increases, the interior angles become larger, making the exterior angles smaller.

EXAMPLE 6–5   Exterior Angles

If a regular polygon has 15 sides, what will be the measure of each of its exterior angles?

◆ **SOLUTION**

The total number of degrees in all the exterior angles will be 360°. Since each side has one corresponding exterior angle, each exterior angle will measure $\dfrac{360°}{15} = 24°$. ◆

6 | Exercises

*For exercises 1–4, use the terms from the **Vocabulary Checklist** to fill in the blanks.*

✔

VOCABULARY CHECKLIST:

triangle	decagon
quadrilateral	n-gon
pentagon	regular polygon
hexagon	diagonal
heptagon	convex polygon
octagon	concave polygon
nonagon	

1. A polygon with six congruent angles would be referred to as a regular __hexagon__ .

2. A concave polygon is one for which a _____ falls outside the polygon.

3. A polygon with 24 sides would be referred to as a(n) ___24___ .

4. A regular polygon is always a(n) _____ .

For exercises 5–8, answer TRUE or FALSE.

T | F 5. An *n*-sided polygon has *n* − 3 diagonals from each vertex, whether it is regular or not.

T | F 6. A polygon cannot be convex and concave at the same time.

T | F 7. Given the number of sides in a regular polygon, the measure of each angle can be found by the formula $A = \dfrac{180°(n-2)}{n}$.

T | F 8. A polygon with six sides is called an octagon whether it is regular or not.

9. Identify whether each of the following figures qualifies as a polygon.

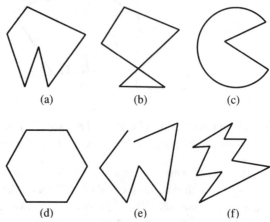

(a)　　　　(b)　　　　(c)

(d)　　　　(e)　　　　(f)

10. Sketch an example of a concave pentagon.

11. Sketch an example of a convex octagon.

12. Sketch an example of a convex heptagon.

13. Sketch an example of a concave quadrilateral.

14. Sketch an example of a concave hexagon.

15. Sketch an example of a convex quadrilateral.

16. What is the one type of polygon that can never be concave?

17. Draw, if possible, a pentagon with exactly two right angles. If it is not possible, explain why.

18. Draw, if possible, a hexagon with exactly two right angles. If it is not possible, explain why.

19. Construct any polygon that is built out of three triangles.

20. Construct any polygon that is built out of five triangles.

21. If a polygon is built out of two triangles that don't overlap, how many sides could it have?

22. If a polygon is built out of three triangles that don't overlap, how many sides could it have?

23. What type of polygon is presented in this photo?

24. What two types of polygons are used to make a soccer ball?

25. There are four types of polygons used in the construction of this greenhouse. What are they? Be as specific as possible.

Exercises 26–41 involve using the theorems about polygons discussed in this section.

26. What is the sum of all the angles in a regular pentagon?

27. What is the measure of each angle of a regular decagon?

28. Each angle of a regular polygon is 172.8°. How many sides does the polygon have?

29. What is the sum of the angles in a regular octagon?

30. What is the sum of the angles in a regular heptagon?

31. How many diagonals could be drawn from one vertex of a nonagon?

32. If the sum of the angles in a polygon is 900°, what type of polygon is it?

33. What is the measure of each angle in a regular quadrilateral?

34. How many diagonals could be drawn from one vertex of a 15-gon?

35. How many sides would a polygon have if the sum of its angles were 4140°?

36. What is the measure of each angle in a regular 1000-gon?

37. What is the sum of the angles in a 45-gon?

38. What will the measure of each exterior angle be for a regular heptagon?

39. How many diagonals could be drawn from one vertex of a 67-gon?

40. What is the sum of the angles in a 516-gon?

41. What will the measure of each exterior angle be in a regular 144-gon?

42. **The Total Number of Diagonals in a Polygon.** In this section we said that the total number of diagonals from one vertex in a polygon is $n - 3$. Let T be the total number of diagonals in a polygon. What formula could be used to find T? Explain.

43. How many diagonals total could be drawn in a regular hexagon?

44. The following diagram is of a regular octagon with all its diagonals drawn. How many diagonals are there total?

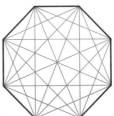

45. How many diagonals total could be drawn in a regular nonagon?

46. How many diagonals total could be drawn in a regular 100-gon?

47. **Geometry and Algebra: Deriving the Number of Sides Theorem.** Start with the formula $A = \dfrac{180(n-2)}{n}$. Solve for n and explain the significance of the new formula.

48. If every angle in a regular polygon is 135°, how many sides does the polygon have?

49. If every angle in a regular polygon is 144°, how many sides does the polygon have?

50. If every angle in a regular polygon is 108°, how many sides does the polygon have?

51. Find the values of x and y shown in the figure below.

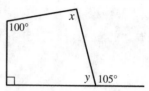

52. Find the value of y.

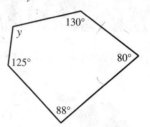

53. Find the value of x.

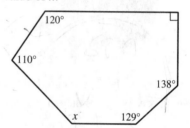

Geometry Project 6 | Dissecting a Regular Hexagon

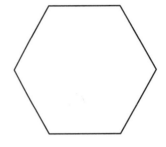

The following steps will take you through partitioning and measuring the parts of a regular hexagon. Your job is to perform each step neatly. Do not use a protractor. We have not provided a figure as a hint for each step because the purpose of the project is to see if you can generate the figure from the instructions; so pay careful attention to written hints and vocabulary.

1. Sketch a diagram of a regular hexagon with side length 4 cm. The angles don't have to be perfect, but use a straightedge to draw the sides.

2. Divide the hexagon into six congruent triangles that share a vertex at the center of the hexagon. This can be done by connecting the diagonals of opposite vertices.

   a. What is the measure of each of the six central angles created?

   b. What types of triangles are these and how do you know? (*Hint:* The line segments used to draw the triangles are all congruent.)

3. For one of the six triangles, draw the angle bisector for the noncongruent angle. Extend the angle bisector so that it ends at the opposite side. This splits this one triangle into two congruent triangles.

   a. How do you know the two triangles are congruent? (*Hint:* You can cite one of the congruence theorems from Section 4.)

   b. This creates two special triangles; what kind of special triangles are they?

4. Use the results from step 3 and techniques for special triangles from Section 4 to label the angle measures and side lengths of the two congruent triangles created in step 3.

Bonus

5. In Section 7: Area and Perimeter, we will see that the amount of space (area) taken up by a triangle is given by $A = \frac{1}{2} \cdot b \cdot h$. The figure to the right gives a reference. What then is the area of one of the triangles from step 3?

6. What is the area of the whole hexagon?

7. Use the same process to derive a generic formula for the area of a regular hexagon with side length x.

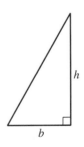

Area and Perimeter

While the Greeks were responsible for a rigorous, systematic approach to geometry, it is generally assumed that the Egyptians were the first to use geometry for practical reasons. The earliest writing on geometry came from the Egyptians in about 1700 B.C. One factor that motivated the Egyptians to use geometry was the desire to divide land between people.

The image in the margin is of a large piece of land divided into several parts by roads. How much land is in each of the sections is an important question that can be answered using the techniques found in this section.

- Perimeter

- Area

- Derived Area Formulas

- More Complex Shapes

■ Perimeter

The **perimeter** of a polygon is the distance around the sides. This can be found by adding the lengths of its sides. Because it represents a distance, perimeter has a linear unit attached, such as feet, meters, centimeters, miles, and so on.

EXAMPLE 7-1   The Perimeter of a Pentagon

Find the perimeter of the pentagon below.

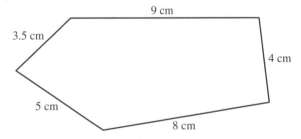

◆ SOLUTION

In geometry, the variable P usually represents perimeter. So, in this case $P = 9$ cm $+ 4$ cm $+ 8$ cm $+ 5$ cm $+ 3.5$ cm $= 29.5$ cm.   ◆

Since a rectangle has opposite sides that are congruent, the perimeter of a rectangle can be expressed by the formula $P = 2w + 2\ell$. Note the illustration in the margin.

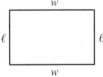

$$P = w + \ell + w + \ell$$
$$= 2w + 2\ell$$

EXAMPLE 7-2   Working with the Rectangle Perimeter Formula

The perimeter of a rectangle is 75.8 feet. If the width of the rectangle is 18.5 feet, what is the length?

◆ SOLUTION

We have been given that $P = 75.8$ ft and $w = 18.5$ ft.

$w = 18.5$ ft

$\ell = ?$ $P = 75.8$ ft $\ell = ?$

$w = 18.5$ ft

If we substitute these values into the formula $P = 2w + 2\ell$, we get the equation $75.8 = 2(18.5) + 2\ell$. Now we can solve for ℓ.

$$75.8 = 2(18.5) + 2\ell$$
$$75.8 = 37 + 2\ell$$
$$38.8 = 2\ell$$
$$19.4 = \ell$$

So the length is 19.4 feet. Finally, we do a quick check to make sure these measurements add to give the correct perimeter: $2(18.5) + 2(19.4) = 37 + 38.8 = 75.8.$ ✓ ◆

In certain cases, you may not know a side length and have to find it before you can find the perimeter, as in this next application.

EXAMPLE 7-3   The Cost of Fencing

A business wants to have their lot fenced in. The lot is pictured below. If fencing costs $13.50 per foot, what will the total cost be?

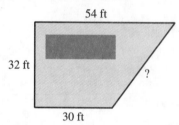

54 ft

32 ft

?

30 ft

◆ SOLUTION

The lot is in the shape of a trapezoid, but we don't know one of the side lengths. To find this length, notice that the lot could be partitioned into a triangle and a rectangle.

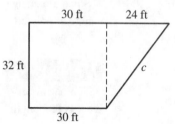

30 ft   24 ft

32 ft

c

30 ft

Thinking of the trapezoid this way makes the missing length the hypotenuse of a right triangle. We have labeled it c and can find its length using the Pythagorean Theorem:

$$c^2 = 24^2 + 32^2$$
$$c^2 = 1600$$
$$c = 40$$

So the entire perimeter is $P = 54 + 40 + 30 + 32 = 156$ ft. Since the cost per foot is $13.50, we multiply $13.50 \cdot 156 = 2106$. So the total cost would be $2106.   ◆

■ Area

The **area** of any two-dimensional object measures the amount of flat space it takes up. To take such a measurement, you would need to use square units. One square unit is the amount of space taken up by a square that has sides of length 1 unit. Here are two examples.

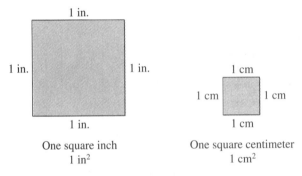

One square inch
1 in^2

One square centimeter
1 cm^2

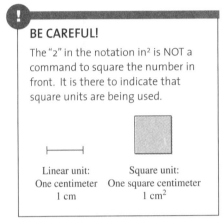

BE CAREFUL!

The "2" in the notation in² is NOT a command to square the number in front. It is there to indicate that square units are being used.

Linear unit:
One centimeter
1 cm

Square unit:
One square centimeter
1 cm^2

The total amount of space in a house would be given in square feet (ft^2). If you measure the area of an entire state, you would probably use square miles (mi^2) or square kilometers (km^2). The state of California, for example, is 158,706 mi^2.

To better illustrate the concept of area, suppose we wanted to know how much space is taken up by the object in the margin. One way to accomplish this would be to impose a grid onto the figure, as shown below.

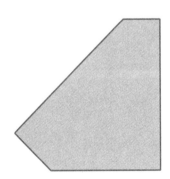

In this case, the units are square centimeters (cm^2). Every square on the grid represents one square centimeter. Now we can just count the total number of square centimeters covered by the object, noting that some of the sections only take up half a square. So the total area of the object is $A = 4 \cdot \frac{1}{2} \text{ cm}^2 + 9 \cdot 1 \text{ cm}^2 = 2 \text{ cm}^2 + 9 \text{ cm}^2 = 11 \text{ cm}^2$.

This technique is most effective in finding the area of a rectangle, and even leads to a general formula for finding the area of a rectangle given its length and width. Consider the rectangle below. We have divided the length and width into 1-inch pieces. Notice that this creates squares that have side lengths of 1 inch, so each square is 1 square inch (1 in^2). Notice also that each of the eight vertical strips has six squares, so the total area is $8 \cdot 6 = 48 \text{ in}^2$. This is true in general. That is, to find the area of a rectangle, we multiply its length and width.

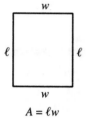

$A = \ell w$

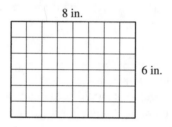

8 in.

6 in.

While this technique illustrates the concept of area well, it will not work for all objects.

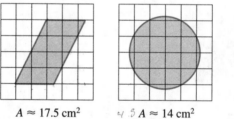

$A \approx 17.5 \text{ cm}^2$ $A \approx 14 \text{ cm}^2$ $A \approx 10 \text{ cm}^2$

Notice in these figures, having the grid allows us only to *approximate* the area of each object and not get an exact value. Fortunately, formulas can be derived for the areas of the popular shapes that we use in this book. As we will see, the justification for these formulas is based on the formula for the area of a rectangle: $A = \ell \cdot w$.

■ Derived Area Formulas

Let's run through the derivation of the formulas for the area of some popular shapes.

Parallelogram: $A = b \cdot h$

Take a parallelogram and cut it into two parts along its perpendicular height (labeled h). If you reorient the two resulting pieces, you can create a rectangle that looks like the second figure below. This shows that the area of a parallelogram is $A = b \cdot h$, where b is the length of its base and h is its perpendicular height.

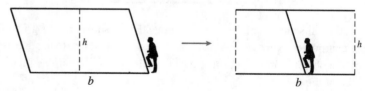

Triangle: $A = \frac{1}{2} \cdot b \cdot h$

If we make a copy of the triangle, we can reorient the copy so that the two make a parallelogram. As shown, the area of a parallelogram is $A =$ base $\cdot$ height. But we only want half of that since our parallelogram is made out of two congruent triangles. So, for a triangle: $A = \frac{1}{2} \cdot b \cdot h$.

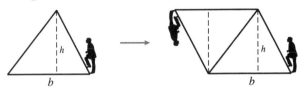

Trapezoid: $A = \frac{1}{2}h(B + b)$

If we take a copy of the trapezoid and rearrange it, we can make a parallelogram. While this is indeed a parallelogram, its base and height are actually $B + b$ and h, respectively. So the entire area of the figure to the right is $A = h \cdot (B + b)$. We only want half this area, so for a trapezoid: $A = \frac{1}{2}h(B + b)$.

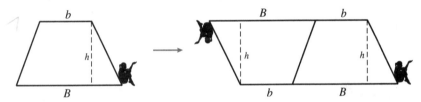

The area formulas for a rectangle, parallelogram, triangle, and trapezoid should be committed to memory.

EXAMPLE 7–4   The Area of a Parallelogram

Find the perimeter and area of the following parallelogram.

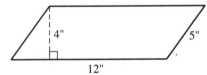

◆ **SOLUTION**

The trick here is knowing which measurement to use for which calculation and knowing which units are appropriate in each case.

- The perimeter is the distance around the object, so it doesn't involve the perpendicular height at all: $P = 12 + 5 + 12 + 5 = 34$ in. Notice the unit is inches.

- The formula for the area of a parallelogram is $A = bh$, where h is the perpendicular height, so $b = 12$ and $h = 4$. So $A = bh = 12 \cdot 4 = 48$ in^2. Notice the unit is square inches.   ◆

EXAMPLE 7-5   The Perimeter and Area of a Triangle

Find the perimeter and area of the following triangle.

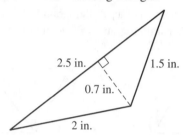

◆ SOLUTION

• Finding the perimeter is straightforward:

$$P = 1.5 + 2.5 + 2 = 6 \text{ in.}$$

• For the area, we note that even though the perpendicular height is not oriented straight up and down the way we are used to seeing it, it is still the perpendicular height. Since the perpendicular height is always perpendicular to the base, $h = 0.7$ in. and $b = 2.5$ in. So, for the area, $A = \frac{1}{2} \cdot 2.5 \cdot 0.7 = 0.875 \text{ in}^2$.   ◆

EXAMPLE 7-6   Working with an Area Formula

Find the smaller base of the trapezoid, given that the area is 45 ft².

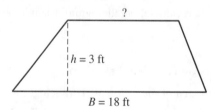

◆ SOLUTION

Given the area of 45 square feet and the values we already know, the area formula $A = \frac{1}{2}h(B + b)$ becomes

$$
\begin{aligned}
45 &= \frac{1}{2} \cdot 3 \cdot (18 + b) & \\
90 &= 3(18 + b) & \textit{clear fractions} \\
30 &= 18 + b & \textit{divide by 3} \\
b &= 12 & \textit{subtract 18}
\end{aligned}
$$

So the smaller base is 12 feet. We can make a quick substitution of this value back into the area formula to make sure we get an area of 45 square feet:

$$A = \frac{1}{2}h(B + b) = \frac{1}{2} \cdot 3 \cdot (18 + 12) = \frac{1}{2} \cdot 3 \cdot 30 = \frac{1}{2} \cdot 90 = 45✓ \qquad ◆$$

■ More Complex Shapes

Some figures are composed of different shapes joined together. The technique used in such situations is to separate the object into smaller pieces whose areas are easy to find. Consider the figure below and suppose we want to find the area of the shaded region. We show the figure before it is divided and then again after it is divided, the idea being that

Shaded Area = Area of a Triangle + Area of a Trapezoid − Area of a Rectangle.

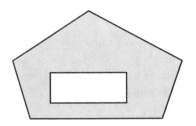
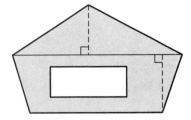

This technique is particularly useful in applications, as we will see in the next example.

EXAMPLE 7-7   Bricking a Patio

A mason is being hired to cover the patio area below with brick. The rectangle in the center represents the customer's grill area, which doesn't need to be covered. If the mason charges $7.50 per square foot for labor and materials, what will the total charge be?

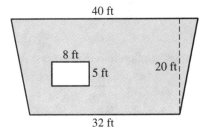

◆ **SOLUTION**

To find the charge, we have to find the total area to be covered. We can use the equation

$$\text{Total Area} = \text{Area of Trapezoid} - \text{Area of Rectangle}$$
$$\text{Total Area} = \frac{1}{2} \cdot h(B + b) - \ell \cdot w$$
$$= \frac{1}{2} \cdot 20(40 + 32) - 8 \cdot 5$$
$$= 10(72) - 40 = 720 - 40 = 680$$

A total of 680 square feet is to be covered with brick. At $7.50 per square foot, the total charge will be $7.50 · 680 = $5100.   ◆

EXAMPLE 7-8   The Area of a Complex Shape

Find the shaded area in the following figure.

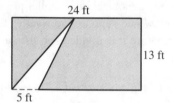

◆ **SOLUTION**

The figure shows a rectangle that needs to have the area of a triangle subtracted out of it. Notice that the length of the perpendicular height of the triangle is not labeled explicitly. But since the rectangle is 13 feet high and the triangle extends the length of it, we can use a perpendicular height of 13 feet for the triangle. So, for the area:

$$A = \ell \cdot w - \frac{1}{2} \cdot b \cdot h = 24 \cdot 13 - \frac{1}{2} \cdot 5 \cdot 13 = 312 - 32.5 = 279.5 \text{ ft}^2. \quad ◆$$

EXAMPLE 7-9   Finding Missing Values

Find the area and perimeter of the figure below.

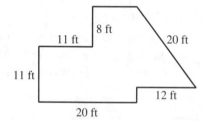

◆ **SOLUTION**

First, we divide the building into familiar geometric shapes.

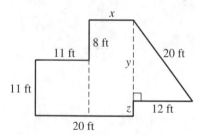

Notice, however, we don't know a few of the lengths, so we have labeled them with variables. We begin by finding these missing values.

• Find x: Since the total length is 20 feet and the left rectangle is 11 feet wide, we have

$$11 + x = 20$$
$$x = 9$$

• Find y: The length y is the height of our triangular piece and is a little more challenging to find. Notice that y is the only missing piece of a right triangle. By the Pythagorean Theorem:

$$y^2 + 12^2 = 20^2$$
$$y^2 = 256$$
$$y = 16$$

- Find z: Knowing y makes z easy to find since, from looking at the left sides, we see the height of the figure is $11 + 8 = 19$ ft.

$$y + z = 19$$
$$16 + z = 19$$
$$z = 3 \text{ ft}$$

So the figure can now be redrawn with all the needed measurements.

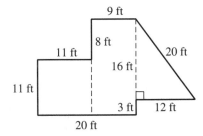

To find the perimeter, we add the side lengths that make up the original figure:

$$11 + 11 + 8 + 9 + 20 + 12 + 3 + 20 = 94$$

So the perimeter is 94 feet.

For area, we use the equation

Total Area = Area of Square + Area of Rectangle + Area of Triangle

Total Area $= s^2 + \ell w + \frac{1}{2}bh$

$$= 11^2 + 19 \cdot 9 + \frac{1}{2} \cdot 16 \cdot 12$$
$$= 121 + 171 + 96 = 388$$

Therefore, the total area is 388 ft^2.   ◆

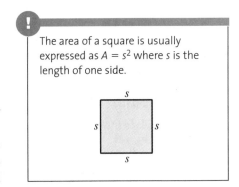

The area of a square is usually expressed as $A = s^2$ where s is the length of one side.

We redrew the figure two times in this problem to help illustrate the process. When you do these problems in the homework, you may not even have to draw the figure over at all; just so long as you can imagine it being divided in a way that lets you find the area.

Section 7 | Exercises

For exercises 1–4, use the terms from the **Vocabulary Checklist** *to fill in the blanks.*

✓

VOCABULARY CHECKLIST:

linear unit   perimeter

square unit   area

1. The total amount of space taken up by a house would be measured using a(n) ___square unit___

2. If I say the distance around my rectangular yard is 176 feet, I have given the _____ of the yard.

3. The ___area___ of the state of Texas is 268,581 square miles.

4. The kilometer is an example of a _____.

For exercises 5–8, answer TRUE or FALSE.

T I F̲   5. The following statement makes sense: "The total length of a football field is 100 square yards."

T̲ F   6. Any two-dimensional object can have a corresponding formula that will give its area.

T I F̲   7. To find the perimeter of an object, you just take the product of its sides.

T̲ I F   8. The following statement makes sense: "The total area of my house is 1540 square feet."

9. **Using a Grid to Approximate Area.** Use the grid to help approximate the shaded area inside. Assume the grid measures the units given.

Square centimeters

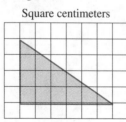

10. **Using a Grid to Approximate Area.** Use the grid to help approximate the shaded area inside. Assume the grid measures the units given.

Square feet

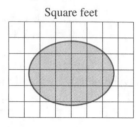

11. **Using a Grid to Approximate Area.** Use the grid to help approximate the shaded area inside. Assume the grid measures the units given.

Square kilometers

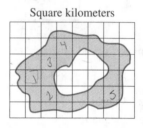

12. Find the area and perimeter of the square.

20 in.

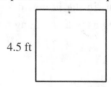

13. Find the area and perimeter of the square.

4.5 ft

14. Find the area and perimeter of the square.

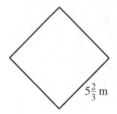

$5\frac{2}{3}$ m

15. Find the area and perimeter of the rectangle.

25 ft

3 ft

16. Find the area and perimeter of the rectangle.

3.5 cm

7.2 cm

17. Find the area and perimeter of the rectangle.

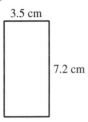

$3\frac{1}{2}$ m

$6\frac{2}{3}$ m

18. Find the area and perimeter.

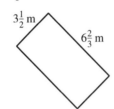

17'

8'

10'

21'

19. Find the area and perimeter.

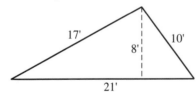

5.9"

4.6"

3.8"

3.4"

20. Find the area and perimeter.

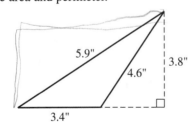

10 in.

6 in.

5 in.

8 in.

21. Find the area and perimeter.

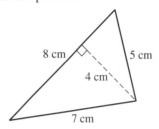

8 cm

5 cm

4 cm

7 cm

22. Find the area and perimeter of the parallelogram.

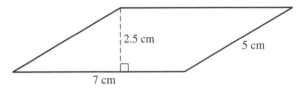

2.5 cm

5 cm

7 cm

23. Find the area and perimeter of the parallelogram.

2.8 cm

2.4 cm

6.7 cm

24. Find the area and perimeter of the parallelogram.

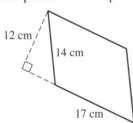

12 cm

14 cm

17 cm

25. Find the area and perimeter of the trapezoid.

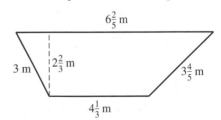

$6\frac{2}{5}$ m

3 m

$2\frac{2}{3}$ m

$3\frac{4}{5}$ m

$4\frac{1}{3}$ m

26. Find the area of the trapezoid.

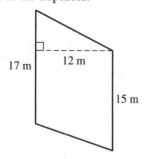

27. Find the area and perimeter of the trapezoid.

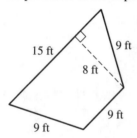

28. Find the area and perimeter. Assume all the angles are right angles.

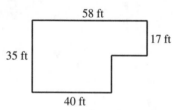

29. Find the area.

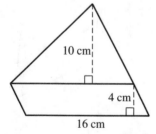

30. Find the area and perimeter of the trapezoid.

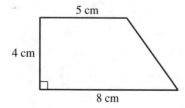

31. Find the area and perimeter.

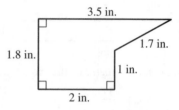

32. Find the area.

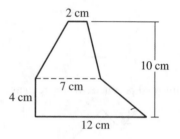

33. Find the area.

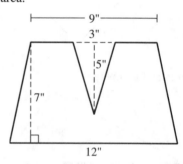

34. Find the shaded area.

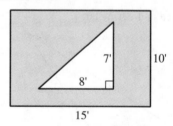

35. Find the shaded area.

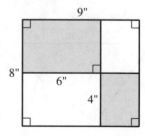

Below are maps of North Dakota, Tennessee, Nevada, and Utah. Notice the scales on the maps that represents either 80 miles or 100 miles. Use an appropriate formula to approximate each state's area and perimeter.

36. **North Dakota**

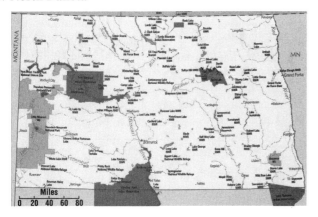

37. **Tennessee**

38. **Nevada**

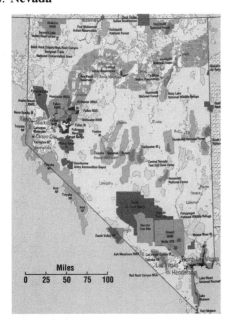

39. **Utah**

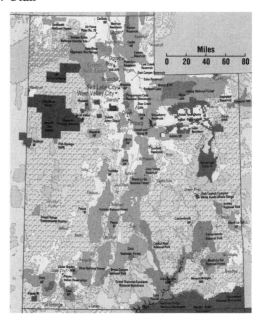

40. A triangle has a base of 4.8 inches. Its height is three times as long as its base. What is the area of the triangle?

41. A certain rectangle has a perimeter of 32 meters. Its length is 5 meters less than twice its width. What are the dimensions of the rectangle?

42. A trapezoid has a height of 6 inches. Its smaller base is 3 inches more than twice its height. Its larger base is four times its height. Find the area of the trapezoid.

43. A certain isosceles triangle has a perimeter of 42 inches. The unequal side is one-third the length of the equal sides' length. What is the length of each side?

44. **Isosceles Triangles.** In an isosceles triangle the angles opposite congruent sides are equal. Furthermore, the height of the triangle drawn from the vertex opposite the unequal side bisects the unequal side. This information is summarized in the figure below.

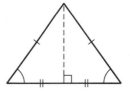

If an isosceles triangle has two sides that are 5 centimeters and a base that is 6 centimeters, find the height of the triangle and its area. (*Hint:* You will need the Pythagorean Theorem.)

45. Sketch any rectangle that has an area of 36 square centimeters.

46. Sketch and label any parallelogram that has an area of 24 square inches.

47. Sketch any rectangle that has a perimeter of 24 inches.

48. Sketch any rectangle that has an area of 72 square meters.

49. Sketch any parallelogram that has a perimeter of 25 feet.

50. Sketch any trapezoid that has an area of 54 square inches.

51. **Pricing Hardwood Floors.** The Smiths are considering getting hardwood floors in their living room. The room is rectangular with dimensions 25 feet by 32 feet. If hardwood floors cost $12.25 per square foot, how much will it cost the Smiths to have their living room covered?

52. **The Cost of a Billboard.** A standard billboard has dimensions 12 feet by 8 feet. A graphics company charges $9.50 per square foot to print the billboard posters. What is the typical charge for a billboard?

53. **Custom Framing.** A custom frame shop charges $2.75 per square inch to frame a picture. If Holly wants a rectangular picture with dimensions 13 inches by 9 inches framed, how much will it cost?

54. **Artistic Murals.** Vincent charges $75 per square foot to do murals. A client wants a mural drawn in an area the shape of an isosceles triangle with a base of 8 feet and a height of 10 feet. What will Vincent charge to paint the mural?

55. **Fertilizing a Garden.** A rectangular garden of dimensions 70 feet by 95 feet needs to be completely covered with fertilizer. If one bag of fertilizer costs $15 and covers 350 square feet, how many bags will be needed and what will the total cost be?

56. **Land for a Park.** The government wants to set aside a rectangular piece of land as a national park. The land is on a thin strip with parallel boundary lines that are a distance of 3 miles apart. If the idea is to set aside 35 square miles of land, how many miles long should the strip be?

57. **Establishing a Garden.** Holly has established an area with the following dimensions as a flower garden.

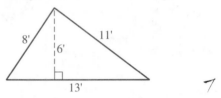

A small rock retaining wall around the perimeter will cost $5 per linear foot. Fertilizer will cost $2 per square foot. How much will it cost to build the retaining wall and fertilize the garden?

58. **Fencing an Area of Land.** If it costs $7.50 per linear foot for fencing, how much will it cost to fence in the following yard?

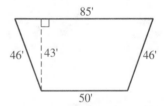

59. **Laying a Driveway.** A company charges $55 per square yard to pour, set, and smooth concrete. Roger has the following design in mind for a driveway. What will the company charge him to lay concrete for the driveway?

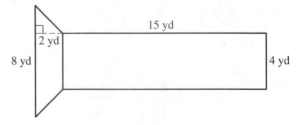

60. **Carpeting a Room.** Given that carpet costs $3.50 per square foot, how much would it cost to carpet the following area?

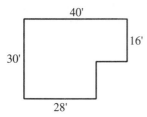

61. **Landscaping.** A well-renowned landscaper charges $8 per square yard of green in the area he is hired to develop, and you are wondering if you can afford him. Your yard has the shape shown next. The rectangles shown in the yard represent your house and a utility building in back. What will the landscaper charge you for a visit in which he cares for the whole yard?

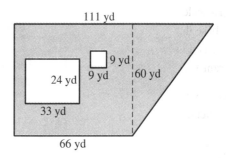

62. **Finding a Perimeter.** Suppose you want to build a fence around the yard illustrated in exercise 61. Find the perimeter of the entire yard. Express your answer in yards. (*Hint:* You will need to use the Pythagorean Theorem.)

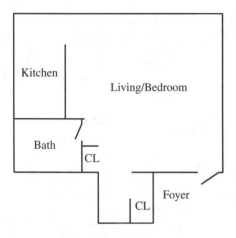

Geometry Project 7 | Your Dream House

Congratulations! You are about to design your dream home! In the process, you will keep track of the measurements involved with this type of plan.

Here are some guidelines for the project:

1. The example in the margin sure is boring. A dream house should at least involve triangles and trapezoids; maybe even circles (see Section 8). Make sure to use at least two shapes other than rectangles.

2. Your house should have *at least* the following:
   - a kitchen
   - two bedrooms
   - one bathroom
   - two closets

   Then, feel free to give yourself a recreation room, a three-car garage, an indoor basketball court; whatever you want.

3. The length should all be to scale. That is, you could use 1 inch to represent 5 feet. Or, if you know you want a really big house, let 1 centimeter represent 5 feet. However you scale the work, make sure the measurements are accurate and proportional.

4. Label the length of every wall.

5. For each room, use a formula from this section (or Section 8 if you feel ambitious) to find and label its area in square feet.

6. To the side of the blueprint, include the following information:
   - the total perimeter of the house
   - the total area of the house
   - the total cost of having your house built (use $90 per square foot)

Circles

The circle represents a definite shift in our study. All the objects we have studied so far share a characteristic that can be summed up in one word: "straightness." A circle's lack of "straightness" is its defining characteristic.

The oldest wheel ever discovered in an archeological excavation dates back to about 6,000 years and was found in what was ancient Mesopotamia (now parts of Iraq, Syria, and Turkey.

■ Definitions and Terminology

A **circle** is the set of all points *in a plane* that lie at a specific distance from a given point called the **center.** All line segments that have one endpoint at that center point and the other on the circle are the same length. Each of these segments is called a **radius** (*r*) of the circle. The plural form of the word radius is radii, so we say that all radii of the same circle have the same length. The word *radius* also refers to the length of a radius.

A line segment drawn from one side of the circle to the other that passes through the center of the circle is called a **diameter** (*d*). All diameters of the same circle have the same length. The word *diameter* also refers to the length of such a line segment. The length of the diameter is two times the length of the radius, so $d = 2 \cdot r$.

Here are some other terms we will use in exploring the properties of circles, along with a figure that illustrates all the terms:

- A **chord** is any line segment with its endpoints on the circle. So every diameter is a chord, but not all chords are diameters.

- A **tangent line** is a line that touches the circle at only one point. A **tangent segment** or **tangent ray** is one that, if extended, would create a tangent line.

- A **secant** is a line that touches the circle twice.

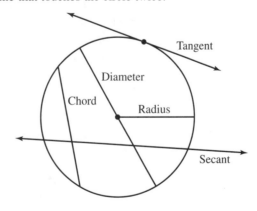

EXAMPLE 8-1   Using Terminology

Consider the circle below and do the exercises that follow.

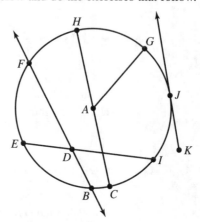

a. How many radii are shown? Name them.

b. Name a diameter for the circle.

c. Name a chord for the circle.

d. Name a secant for the circle.

e. Name a tangent for the circle and indicate what type of tangent it is.

f. If $HC = 18$, what is the radius of the circle?

g. If $AG = 4.5$, what is the diameter of the circle?

h. If $\overline{HG}$ were drawn, would it be a chord, secant, or tangent?

◆ SOLUTION

a. There are three radii given: $\overline{AG}$, $\overline{AC}$, and $\overline{AH}$. Notice that two of these form a diameter together.

b. The diameter is given by $\overline{HC}$.

c. The chord is given by $\overline{EI}$.

d. The secant is given by $\overleftrightarrow{FB}$.

e. The tangent is given by the ray $\overrightarrow{KJ}$, so it is a tangent ray.

f. Since $\overline{HC}$ is a diameter, the radius is half its length, so the radius $r = 9$.

g. Since $\overline{AG}$ is a radius, the diameter is twice its length, so the diameter $d = 9$.

h. If $\overline{HG}$ were drawn it would be a chord since it would be a line segment that has its endpoints on the circle.   ◆

There are several theorems pertaining to these objects that are easy to prove. Here are some examples:

1. **Theorem:** *A tangent to a circle is perpendicular to exactly one radius and exactly one diameter.*

2. **Theorem:** *If two distinct tangent segments share an endpoint and each has their second endpoint on the circle, then they are congruent.*

3. **Theorem:** *A radius that is perpendicular to a chord bisects that chord.*

EXAMPLE 8-2   Illustrating Theorems

Sketch figures that illustrate the first two theorems mentioned above.

◆ **SOLUTION**

1. **Theorem:** *A tangent to a circle is perpendicular to exactly one radius and exactly one diameter.*

This theorem involves a tangent, a radius, and a diameter, so the illustration must involve these objects. The tangent could be a line, ray, or segment, but we'll use a line to illustrate.

2. **Theorem:** *If two distinct tangent segments share an endpoint and each has their second endpoint on the circle, then they are congruent.*

We begin by drawing a point on the exterior of the circle and forming two tangent segments to the circle from that point. And since the theorem is saying that the two resulting tangent segments are congruent, we mark them with tick marks.

The illustration of the third theorem is left to you as an exercise.   ◆

While we present and explore this information in a very abstract way, circles and the objects related to them are practical for a number of sciences, including engineering and astronomy. The figure below gives a geometric explanation for the occurrence of a major and minor lunar eclipse. Notice that the figure involves circle tangent rays.

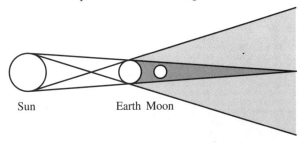

Sun   Earth Moon

Archimedes is the Greek mathematician attributed with the first close approximation of *pi* (π) using regular polygons.

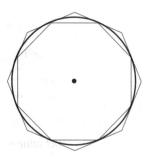

The value of π to eight decimal places is π = 3.14159265.

■ Circumference and Area

It is often useful to know the distance around a circle. The concept behind this is the same as for the perimeter of a polygon. But a circle is not a polygon, so we use the term *circumference* instead of perimeter: the **circumference** of a circle is the total distance around the circle.

The question of how the radius or diameter and the circumference of a circle are related was a puzzle for thousands of years. The Greek mathematician Archimedes (287 B.C.−212 B.C.) tried to find out how they are related by drawing polygons inside and outside a circle and finding the perimeter of the polygons. When he used polygons with more and more sides, the perimeters of the polygons became closer to the circumference of the circle.

Eventually, it was discovered that if we can accurately measure both the circumference C and the diameter d of a circle, then C divided by d always results in the same number. This number is called π (spelled pi). π is an irrational number, meaning it never terminates or repeats. π is usually approximated to two decimal places ($\pi \approx 3.14$) or as the fraction $\frac{22}{7}$. So we have $\frac{C}{d} = \pi$, or $C = \pi \cdot d$. Since the diameter is twice as long as the radius, we can also say $C = 2\pi r$. This is the most commonly used formula.

EXAMPLE 8-3   The Circumference of a Circle

If a circle has a radius of 12 inches, find its circumference. Give an answer that includes π, then use $\pi \approx 3.14$ to get an approximation to the nearest hundredth.

◆ SOLUTION

The circumference can be calculated as

$$C = 2\pi(12) = 24\pi \approx 24(3.14) \text{ inches} = 75.36 \text{ inches}.$$

The answer 24π is called the exact form, and 75.36 is an approximate answer.   ◆

The **area** enclosed by a circle is another important quantity to consider. Archimedes used his polygon method on this problem too and discovered that the area of the inside or outside polygons was a little more than 3 times the square of the radius of the circle. The exact formula is $A = \pi r^2$.

The best way to explore this formula is to notice that if we slice a circle into several pieces it can be approximated by a rectangle.

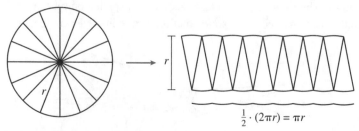

$$\tfrac{1}{2} \cdot (2\pi r) = \pi r$$

In this case the formula $A = \ell \cdot w$ becomes $A = r \cdot \pi r = \pi r^2$. And while the figure above is not a true rectangle, it can be shown that if we slice the circle into even smaller pieces the figure gets closer to a rectangle.

In case you have a hard time believing this approach, there is an easy way to show that it is reasonable. To start, we draw a circle of radius r, circumscribe a square around it, and draw four radii to make four squares total. The area of the circle is obviously less than the total area of the four squares, which is $r^2 + r^2 + r^2 + r^2 = 4r^2$. Since it looks like the four white areas make a total of about one square, then the area of the circle would be close to $4r^2 - r^2 = 3r^2$. Since $\pi \approx 3.14$, the formula $A = \pi r^2$ is at least close. But, in fact, it is exact.

EXAMPLE 8-4   The Circumference and Area of a Circle

If a circle has a radius of 10 meters, find its circumference and area. Use $\pi \approx 3.14$.

◆ SOLUTION

Circumference: $C = 2\pi r = 2\pi (10) = 20\pi \approx 62.8$ m
Area: $A = \pi r^2 = \pi(10)^2 = 100\pi \approx 314$ m^2

EXAMPLE 8-5   Interesting Result for Agriculture

An agriculture specialist has a square plot of land to plant on. The square plot is 72 yards on each side, but the crop will be planted in a circle. Which of the following will provide more crop area?

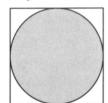
Plant one circular crop that is as large as possible.

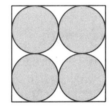
Plant four smaller equally sized crop circles as large as possible.

The ancient Egyptians were content when measurements were *close* to being correct. They would, for example, have no problem with using a square to approximate the area of a circle.

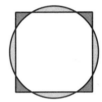

The reasoning here is that the dark and light shaded regions balance each other out.

The Greeks were more concerned with proving the *exact* value of areas of geometric objects.

◆ SOLUTION

Let's find the total area in each situation.

• One large circle: Since the square has a side length of 72 yards, the largest circle that could go in it would have a radius of 36 yards. This means the total area would be

$$A = \pi r^2 = \pi(36)^2 \approx 3.14 \cdot 1296 = 4069.44 \text{ yd}^2.$$

• Four smaller equal circles: The length of the square is two diameters of the circles. This means the diameter of one circle is 36 yards. So the radius of one circle is 18 yards. The area of one of the small circles is

$$A = \pi(18)^2 \approx 3.14 \cdot 324 = 1017.36 \text{ yd}^2.$$

Since there are four such circles, we multiply this area by 4:

$$1017.36 \text{ yd}^2 \times 4 = 4069.44 \text{ yd}^2.$$

So either decision produces the same area.   ◆

EXAMPLE 8-6   The Perimeter of a Complex Shape

A **semicircle** is half of a circle. The following figure contains a semicircle. Find the area and perimeter. Use $\pi \approx 3.14$.

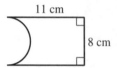

♦ SOLUTION

This example involves what looks like a rectangle, except that one side has been replaced by a semicircle.

- Perimeter: Notice that the diameter of the circle is the length of the rectangle, or 8 cm. So the perimeter can be found by adding the three sides of the rectangle and half the circumference of the circle:

$$P = 11 + 8 + 11 + \frac{1}{2} \cdot 2 \cdot \pi \cdot 4 \approx 30 + 12.56 = 42.56 \text{ cm}$$

- Area: We can find the area of the rectangle and then subtract out the area of the semicircle, which is half the area of a normal circle:

$$A = \ell \cdot w - \frac{1}{2} \cdot \pi \cdot r^2 \approx 8 \cdot 11 - 0.5 \cdot 3.14 \cdot 4^2 = 88 - 25.12 = 62.88 \text{ cm}^2 \quad ♦$$

■ Sectors and Arc Length

A **sector** of a circle is a region enclosed between two radii. An **arc** is a part of the actual circle between two points on the circle. The angle formed by the two radii is called a **central angle.** The area of a sector depends on the angle formed by the two radii. If the radii happen to meet in a right angle or 90°, then the area of the sector is one-fourth of the area of the circle, since 90° corresponds to one-fourth of a complete circle: 360°. If the radii create a 60° angle, then six such sectors would fit inside the circle, so the area of each is one-sixth of the area of the circle. This idea of proportionality leads to a general formula for the area of a sector. We can also use the same idea to find the length of the corresponding arc. We summarize the information and results below.

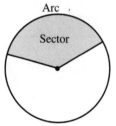

      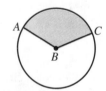

Here, the shaded area is one-fourth the total area, or $A = \frac{1}{4}\pi r^2$.

Here, the shaded area is one-sixth the total area, or $A = \frac{1}{6}\pi r^2$.

In general, the shaded area in a *sector* is
$$A = \frac{m\angle ABC}{360°} \cdot \pi r^2$$

The length of the arc would be one-fourth the total circumference, or $s = \frac{1}{4} \cdot 2\pi r$.

The length of the arc would be one-sixth the total circumference, or $s = \frac{1}{6} \cdot 2\pi r$.

In general, the length of the *arc* is
$$s = \frac{m\angle ABC}{360°} \cdot 2\pi r$$

EXAMPLE 8-7   Finding the Area of a Sector and Arc Length

A circle has a radius of 12 cm. Suppose that a sector of the circle is formed by a central angle of 40°. Find the area of the sector and the arc length. Use $\pi \approx 3.14$ and round to the nearest hundredth.

◆ SOLUTION

The area of the sector is $A = \dfrac{40°}{360°} \cdot \pi \cdot 12^2 = \dfrac{1}{9} \cdot \pi \cdot 144 = 16\pi \text{ cm}^2 \approx 50.24 \text{ cm}^2$.

The length of the arc is $s = \dfrac{40°}{360°} \cdot 2 \cdot \pi \cdot 12 = \dfrac{1}{9} \cdot \pi \cdot 24 = \dfrac{8}{3}\pi \text{ cm} \approx 8.37 \text{ cm}$.◆

EXAMPLE 8-8   Making Snow

A snow maker/blower used by a ski resort fans back and forth throwing snow over a distance of 70 feet. If the blower sweeps out an angle that is 80°, how much area is covered by the blower?

◆ SOLUTION

We use the formula for the area of a sector.

$$A = \frac{m\angle ACB}{360°} \cdot \text{Area of Circle} = \frac{80}{360} \cdot \pi r^2 \approx \frac{2}{9} \cdot 3.14 \cdot 70^2 = 3419.11 \text{ ft}^2$$

So the snow blower covers a total of 3419.11 ft². ◆

8 | Exercises

*For exercises 1–6, use the terms from the **Vocabulary Checklist** to fill in the blanks.*

✓

VOCABULARY CHECKLIST:

circle	pi (π)
radius	circumference
diameter	semicircle
tangent	sector
secant	arc
chord	central angle

1. A part of a circle is called a(n) ___arc___.

2. For a circle, the idea of perimeter (total distance around) is expressed by the term _____.

3. Drawing a line segment from the center of a circle to a point on the circle creates a(n) ___radius___.

4. Two radii and the arc that connects their endpoints create an area called a(n) _____.

5. A(n) ___tangent___ can be a line segment, ray, or line, as long as its corresponding line touches the circle only once.

6. A chord that passes through the center of a circle is called a _____.

For exercises 7–10, answer TRUE or FALSE.

T | F 7. The number π is exactly 3.14.

T | F 8. While they have the same property with regard to the circle, a secant is a line and a chord is a line segment.

T | F 9. It is possible for a line to be both a secant and a tangent to a circle.

T | F 10. The circumference of a circle is exactly half the diameter.

11. Give the diameter of a circle that has the indicated radius.
    a. 6 feet
    b. $3\frac{2}{5}$ in.
    c. 5.8 miles
    d. $(b + 5)$ cm
    e. $(7x)$ meters

12. Give the radius of a circle with the indicated diameter.
    a. 10 meters
    b. $7\frac{2}{5}$ yd
    c. 16.8 km
    d. $(2x - 8)$ cm
    e. $(7x)$ meters

13. The circle below has its center at point C. What line segment(s) represent a radius and what line segment(s) represent a diameter?

14. The circle below has its center at point A. Without naming them, how many diameters are given in the following circle? How many radii?

15. Without naming them, how many diameters are given in the following circle? How many radii?

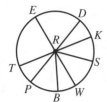

16. Name every object in the following figure and indicate what it is in relation to the circle.

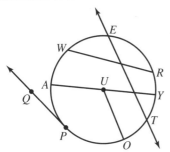

17. Draw a circle and label it so that the following objects are created:

   - radius $\overline{AS}$
   - diameter $\overline{GS}$
   - tangent ray $\overrightarrow{FG}$
   - secant $\overleftrightarrow{RT}$
   - chord $\overline{WQ}$

18. In Example 8-2 we illustrated two of three theorems about various objects pertaining to a circle. Sketch a diagram to illustrate the third theorem mentioned preceding Example 8-2.

19. In the circle below, name each central angle and give its measure. Assume $\overline{HT}$ is a diameter.

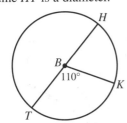

20. In the circle below, name each central angle and give its measure.

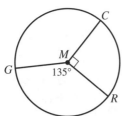

21. Find the area and circumference of the circle. Leave π in your answer.

22. If a circle has a diameter of 6 centimeters, find its circumference and its area. Use $\pi \approx 3.14$ to approximate your answer.

23. If a circle has a radius of 7 feet, find its circumference and its area. Use $\pi \approx \frac{22}{7}$ to approximate your answer.

24. **Geometry Rocks!** A standard compact disc has a diameter of $4\frac{1}{2}$ inches. Find its circumference. Use $\pi \approx \frac{22}{7}$ to approximate your answer.

25. Find the area and circumference of the circle below. Leave π in your answer.

26. **No Fly Zone.** No planes are allowed to fly within three miles of a military base. How many square miles are covered by the "no fly" zone? Use $\pi \approx 3.14$.

27. Find the radius of a circle that has a circumference of 12π meters.

28. Find the area of a circle that has a circumference of 22π inches.

29. Using π ≈ 3.14, find the radius of a circle whose circumference is 94.2 miles.

30. Find the radius of a circle that has an area of 49π in².

31. Find the radius of a circle that has an area of 25π yd².

32. Using π ≈ 3.14, find the radius of a circle whose area is 200.96 square feet.

33. Find the radius of a circle whose circumference in linear units is equal to its area in square units.

34. **A Semicircular Table.** Find the area and perimeter of the semicircular tabletop below, given that it has a radius of 3 feet. Use π ≈ 3.14.

35. **Measurements of a Window.** The window below is made of a rectangle and a semicircle. Find the area of the window. Also, if trim is to be used to frame the interior and exterior of the window as indicated, how much trim will be needed? Use π ≈ 3.14.

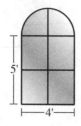

36. Find the area and perimeter of the following figure (assume it is made of semicircles). Use π ≈ 3.14.

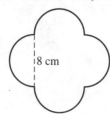

37. Find the area of the shaded region. Use π ≈ 3.14. Assume that the longest side of the triangle is a diameter.

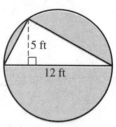

38. Find the area of the shaded region. Use π ≈ 3.14.

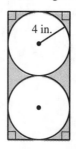

39. Find the area of the shaded region given that the circular part is a semicircle. Use π ≈ 3.14. You will need the Pythagorean Theorem.

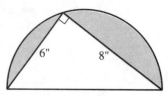

40. Find the area of the shaded region, given that the figure inside the circle is a square. Use π ≈ 3.14 and round to the nearest hundredth if necessary. *Hints:* Remember, the diagonals of a circle form perpendicular line segments. You will need the Pythagorean Theorem.

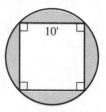

41. **Pizza Time.** Which is the better purchase: one 13-inch diameter pizza for $12.95 or two 9-inch pizzas for $12.49? Use π ≈ 3.14.

42. **Seconds.** If you wanted to make two mini pizzas equivalent in area to one 10-inch diameter pizza, what should the diameter of the mini pizzas be? Round to the nearest inch.

43. **Geometry in Art.** Vincent Burkhead creates mandalas from common everyday pictures. For each of the following mandalas, find the measure of the central angle made by one "slice" of the mandala.

Jasmine

Symphony

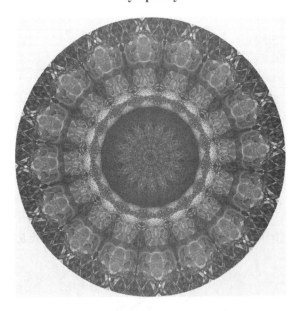

44. A sector of a circle is formed by a central angle of 50°. The radius of the circle is 18 inches. What is the area of the sector? Use $\pi \approx 3.14$.

45. Suppose an arc of a circle corresponds to a central angle of 70°. The diameter of the circle is 3 feet. What is the length of the arc? Use $\pi \approx 3.14$.

46. **Sprinkler System.** A sprinkler system rotates 80°, throwing water over a distance of 15 yards. How many square yards of lawn will the sprinkler water? Use $\pi \approx \frac{22}{7}$.

47. If the length of an arc is 12π feet and the radius is 24 feet, find the measure of the central angle.

48. If the length of an arc is 18π meters and the central angle measures 75°, find the radius of the circle.

49. If the length of an arc is 10π cm and the central angle measures 15°, find the length of the radius.

50. If the length of an arc is 15π inches and the radius of the circle is 45 inches, find the measure of the central angle.

51. If the area of a sector of a circle is 20π square inches and the radius $r = 10$ in., find the measure of the central angle.

52. If the area of a sector is 6π m² and the radius is 12 m, find the measure of the central angle.

53. If the area of a sector of a circle is 90π cm² and the measure of the central angle is 100°, find the radius.

54. If the area of a sector of a circle is 288π yd² and the measure of the central angle is 80°, find the radius.

55. **Geometry and Algebra.** The circumference of the circle below is 8π units. Find x.

56. **Geometry and Algebra.** The circumference of the circle below is 132 units. Find t. Use $\pi \approx \frac{22}{7}$.

57. **Geometry and Algebra.** The area of the circle below is 64π square units. Find x.

58. **Geometry and Algebra.** The area of the circle below is 616 square units. Find b. Use $\pi \approx \frac{22}{7}$.

59. **Corner Rug.** This is a corner rug that has the shape of a quarter circle. Find its area and perimeter, given the radius is 3 feet. Use $\pi \approx 3.14$.

60. **Garden Path.** A circular garden, 20 feet in diameter, has a path around it that is three feet wide. Find the area of the path. Use $\pi \approx 3.14$.

61. **Garden Path.** Here is another circular garden surrounded by a path. The area of the big circle is 144π square yards. The radius of the inner circle is $(x + 5)$ yards, and the width of the path is $(x - 3)$ yards. Find the width of the path. Also find the area of the path.

Exercises 62 and 63 require the use of the Pythagorean Theorem in order to find the area.

62. This "ice cream cone" shape is a semicircle on top of an isosceles triangle. Find the area and the perimeter of the shape. Use $\pi \approx 3.14$.

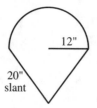

63. This "ice cream cup" shape is a semicircle on top of an isosceles trapezoid. Find the area and perimeter. Use $\pi \approx 3.14$.

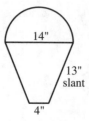

64. **The Orbit of Planet Earth.** The planet Earth is 93 million miles from the sun. How much distance does the Earth travel in four months? Use $\pi \approx 3.14$. *Hint:* If four months is one-third of a year, how many degrees does the Earth travel in that time? Now use the formula for arc length.

65. **Road Trip!** The radius of the moon at its equator is approximately 1080 miles. To the nearest hour, how long would it take to drive around the moon at its equator if you traveled at 60 miles an hour? Use $\pi \approx 3.14$. *Hint: Use distance = rate · time and solve for time.*

66. A circle has a radius of 6 inches. If the circle doubles in radius, how is its circumference affected? How is its area affected?

67. A circle has a radius of 10 feet. If the circle doubles in radius, how is its circumference affected? How is its area affected?

CONSTRUCTION

68. **Ruler and Compass. Concentric circles** are two or more circles that have the same center. Use a compass and/or ruler to create three concentric circles so that the radius of the inner circle is 1 centimeter and each circle after that has a radius that is twice the one before it.

69. **Ruler, Protractor, and Compass.** A **circumscribed circle** is a circle that has been drawn around a polygon such that all the polygon's vertices are points on the circle. The figure here shows a circle *circumscribed* around a square.

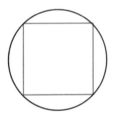

Use a ruler, protractor, and compass to create a circle circumscribed around a square of side length 3 cm. The following theorem will help.

Theorem: *A circle can always be circumscribed around a square. Further, the diagonals of the square are diameters of the circle.*

70. **Straightedge and Compass.** Here is a theorem that addresses the issue of circumscribing a circle around a triangle.

Theorem: *A circle can always be circumscribed around a triangle. Further, if the perpendicular bisectors of each side of a triangle are extended, their point of intersection is the center of the triangle's corresponding circumscribed circle.*

Use a straightedge to draw any acute triangle. Use the theorem above to circumscribe a circle around the triangle. *Hint:* To draw the perpendicular bisectors, you will need the construction technique discussed in the Construction Example in Section 1. According to the theorem above, the perpendicular bisectors should intersect at one point. That point is the center of the triangle's corresponding circumscribed circle.

Repeat the exercise with any obtuse triangle.

Geometry Project 8 | Proofs in Geometry

It's time once again to look at how to prove some of the ideas we have encountered in the last few sections. You may want to review the techniques used for proofs in the projects at the ends of Sections 2 and 3. The proofs you will do in this project involve theorems from other sections of this book. These proofs have been saved until now because we prove a result about tangent line segments to circles

Project Exercises | Writing Geometry Proofs

1. **Theorem:** *The diagonals of a parallelogram bisect each other.*

Here is a proof for this theorem. The statements have been provided. For each statement, provide a justification.

Prove: $AM = MD$, given that the points A, B, C, and D are for a parallelogram.

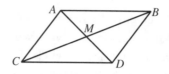

Statement	Justification
$AB = CD$	?
$\overline{AB}$ is parallel to $\overline{CD}$	?
$m\angle MAB = m\angle MDC$ $m\angle MBA = m\angle MCD$	?
$\triangle AMB \cong \triangle CMD$	?
$AM = MD$	?

2. **Theorem:** *Drawing the diagonal of a parallelogram creates two congruent triangles.*

Use the two-column system to prove this theorem. *Strategy hint:* As always, draw a picture to illustrate. You will have to use the *alternate interior angles theorem* and the *ASA theorem.*

3. **Theorem:** *If two distinct tangent segments share an endpoint and each has their second endpoint on the circle, then they are congruent.*

Use the two-column system to prove this theorem. *Setup and strategy hint:* This theorem was stated in Example 8-2 and is illustrated with the following figure. This figure illustrates what you are trying to prove.

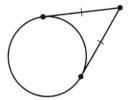

Set up the proof of the theorem using the following figure.

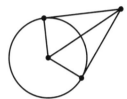

Label points on the figure for reference and try to prove that the figure contains two congruent triangles. This will require a congruence theorem and use of a theorem about circle tangents that appears in the same example as the theorem you are proving (Example 8-2).

Volume and Surface Area

In this section we explore three-dimensional objects, their volume, and their surface area. Here are some examples of volume and surface area that may help you connect with the ideas.

• The newest iPod shuffle can hold up to 240 songs but only has a volume of 0.5 cubic inches.

• The volume of all the Earth's oceans combined is about 1,370,000,000 cubic kilometers.

• The volume of a typical elephant is about 5 cubic meters, while the volume of a typical human is only about 3 cubic feet.

• Because of its spongy nature, the surface area of a human lung is approximately the same as the area of a tennis court.

• The surface area of an apple is usually around 40 square inches.

In geometry, a three-dimensional object is usually referred to as a **solid**. The **volume** of a solid measures the amount of space it takes up. Volume is measured in cubic units. **Surface area** measures the area taken up by the surface of the object. Surface area is measured in square units. Now would be a good time to review the units used for perimeter, area, surface area, and volume. We will use inches in this example, but the same idea can be applied to any linear unit of measure: feet, meters, miles, and so on.

Unit Type	Linear unit	Square unit	Cubic unit
Example (Notation)	Inch (in.)	Square inch (in^2)	Cubic inch (in^3)
What It Looks Like			
What It Measures	Perimeter	Area and surface area	Volume

Here are all the solids we will study in this section.

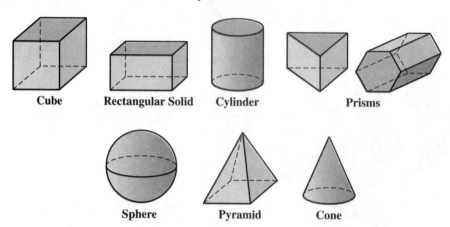

All these figures have formulas that can be used to find their volume and surface area. There are also some concepts we will discuss that are helpful in exploring volume and surface area.

Volume Formulas for Cubes, Rectangular Solids, Cylinders, and Prisms

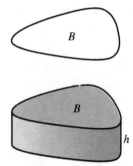

Before we give formulas to calculate volumes, we want to discuss a general concept that will help in understanding the calculations. Suppose we start with a two-dimensional object such as the one in the margin. While it would be difficult to find the area of this object, the area does exist. We will call that area B = the area of the object.

Now suppose we take the same object and create a solid by projecting it upward, so that the result looks like the solid in the margin. We have given the object a height (indicated by the variable h). There is a fascinating and very useful relationship between the two-dimensional object's area and the corresponding solid's volume:

$$\text{Volume} = \text{Area of Base} \cdot \text{Height} = B \cdot h$$

This principle can be applied to any of the following shapes.

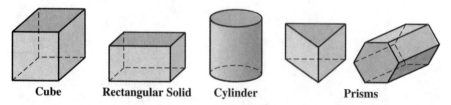

Since all these figures have sides that form right angles with the base, they are called **right.** The cylinder above is a right cylinder, and the prisms are right prisms. Not all solids are right. If the lateral side does not form a right angle, the solid is called **oblique.**

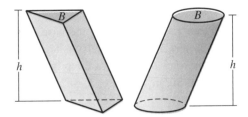

Notice that for these objects, *h* represents the *perpendicular height* of the object. This measurement, and not the actual length of the side, is used to find volume. The good news is that the formula, Volume = Area of Base · Height = $B \cdot h$, still applies to these oblique solids.

Remember, the value of *B* is actually an area. In fact, in this section the variable *B* will always represent the area of the base of a solid. The way the value of *B* is calculated will vary from object to object. For instance, if the base of a solid is a rectangle (as with a box), then the formula for calculating the area of the base would be $B = \ell \cdot w$. If the base is a circle (as with a cylinder), then $B = \pi r^2$.

EXAMPLE 9–1   The Volume of a Triangular Prism

Find the volume of the triangular prism below.

◆ **SOLUTION**

As its name indicates, the base of this prism is a triangle. So, to find the area of the base we use $B = \frac{1}{2}bh = \frac{1}{2} \cdot 8 \cdot 5 = 20$ in².

So the volume of the solid is found by $V = B \cdot h = 20 \cdot 3 = 60$ in³. Note that we switch to cubic inches since we are measuring volume. Also notice that we used the variable h = height twice in the same problem: first as the height of a triangle, then as the height of a solid. Be careful about keeping these straight.   ◆

EXAMPLE 9–2   The Volume of an Oblique Hexagonal Prism

The object below is called a hexagonal prism (because its base is a hexagon). As indicated, its base has an area of 32 in², its perpendicular height is 7.5 in., and its lateral height is 9.3 in. Find its volume.

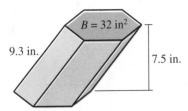

◆ **SOLUTION**

Since we have already been given the area of the base of this solid, the volume can be found quickly by using the formula $V = B \cdot h$. The trick here is that we need to make sure we use the *perpendicular height* of the prism and not the *lateral height* (the actual length of the side). So, in this case, $h = 7.5$. That means the volume is $V = 32 \cdot 7.5 = 240$ in^3. ◆

Here is a more practical example that can be done without the aid of a diagram.

EXAMPLE 9-3   Guitar Picks

A company called Tortoise makes the best bass guitar picks in the world; their shape and flexibility are perfect. Each pick has a base area of 0.9 square inches. If a pick is 0.08 inches thick, how much plastic is used to make one pick? How many picks could be made with 500 cubic inches of plastic?

◆ **SOLUTION**

The value of B is the area of the base, so $B = 0.9$ in^2. The value of h represents the height of the object, so $h = 0.08$ in. So the volume is simply $V = 0.9 \cdot 0.08 = 0.072$ in^3. Note we switched to cubic units since we are dealing with volume.

Now to find out how many picks could be made from 500 cubic inches of plastic. We want to divide 500 cubic inches of plastic into picks that are 0.072 cubic inches each; we have $500 \div 0.072 = 6944.4 \approx 6944$ picks. ◆

As mentioned before, this process leads to simple volume formulas for rectangular solids, cylinders, and prisms. We will summarize these formulas at the end of the section, but you are encouraged to take an intuitive approach in finding these volumes using the method outlined here, and not rely so much on the formulas. For some solids, however, this technique will not work. For the sphere, the pyramid, and the cone, we rely on the formulas summarized below. In these formulas, the variable B still represents the area of the base of the object, and h still represents the perpendicular height. Let's look at the sphere in more detail.

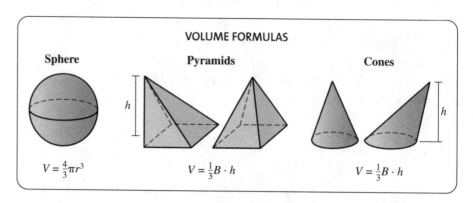

VOLUME FORMULAS

Sphere	**Pyramids**	**Cones**
$V = \frac{4}{3}\pi r^3$	$V = \frac{1}{3}B \cdot h$	$V = \frac{1}{3}B \cdot h$

◼ The Sphere

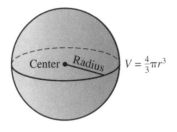

My best math teacher, David Trogdan, used to say, "A sphere is just a circle on steroids." The message in this humorous quote is that the definition and properties of a sphere are identical to those of a circle applied to three dimensions. Recall the definition for a circle: A **circle** is the set of all points *in a given plane* that are a given distance from a center point. A **sphere,** on the other hand, is the set of all points in *any* direction that are a fixed distance from a center point. The distance from the center to the points on the sphere is still called the **radius,** and the center point is still called the **center.**

The volume of a sphere can be found using the formula $V = \frac{4}{3}\pi r^3$, where r is the radius of the sphere. Notice that the only value needed to determine the volume of the sphere is the radius (similar to the area of a circle).

EXAMPLE 9-4   The Volume of the Earth

By around 200 B.C. the Greeks had already used properties of angles to figure out that the radius of the Earth is approximately 4000 miles (although their unit of measure was not miles). What is the approximate volume of the Earth? Use $\pi \approx 3.14$ and express your answer in scientific notation.

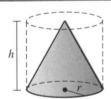

◆ SOLUTION

This problem is a straightforward application of the formula for the volume of a sphere:

$$V = \frac{4}{3}\pi r^3 \approx \frac{4}{3}(3.14)(4000)^3 = 4.19(6.4 \times 10^{10}) = 26.82 \times 10^{10} = 2.682 \times 10^{11} \text{ mi}^3$$

So the volume of the Earth is approximately 2.68×10^{11} cubic miles.   ◆

◼ Cones and Pyramids

Since cones and pyramids are identified by their base and height, each one has a corresponding box, prism, or cylinder that encompasses it. Notice in the formulas for cones and pyramids, the volume is exactly one-third the volume of a prism or cylinder. This can be helpful in working problems if you don't want to have to memorize formulas; you can find the volume of the corresponding cylinder or prism and then multiply by $\frac{1}{3}$.

EXAMPLE 9-5   The Volume of an Egyptian Pyramid

The Great Pyramid of Giza in Egypt is one of the Seven Wonders of the World. Its base is a square with an impressive side length of 230 meters. The height of the pyramid is 150 meters. What is its volume?

◆ SOLUTION

The base is a square, so its area can be found by $B = s^2 = 230^2 = 52{,}900$ m². Now we use the formula given earlier: $V = \frac{1}{3} \cdot B \cdot h = \frac{1}{3} \cdot 52{,}900 \cdot 150 = 2{,}645{,}000$. The volume is 2,645,000 m³. To give you an idea of how large that is, the volume of the Empire State Building is just over 1 million cubic meters.   ◆

The relationship between these two objects is always the same, and can be expressed as follows:

$$V_{cone} = \frac{1}{3} \cdot V_{cylinder}$$

This relationship is also true for a pyramid and the corresponding prism.

EXAMPLE 9–6   The Volume of an Oblique Cone

Find the volume of the object below.

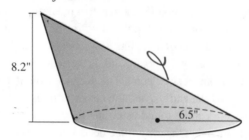

8.2"

6.5"

SUMMARY OF VOLUME FORMULAS	
OBJECT	FORMULA FOR VOLUME
Cube	$V = s^3$
Rectangular Solid	$V = \ell \cdot w \cdot h$
Cylinder	$V = B \cdot h = \pi r^2 \cdot h$
Cone	$V = \frac{1}{3}B \cdot h = \frac{1}{3}\pi r^2 h$
Prism	$V = B \cdot h$
Pyramid	$V = \frac{1}{3}B \cdot h$
Sphere	$V = \frac{4}{3}\pi r^3$

◆ **SOLUTION**

The base of this cone is a circle, so the area of the base can be found using the formula $B = \pi r^2 \approx 3.14 \cdot 42.25 = 132.67$ in². Note that we used $\pi \approx 3.14$, and rounded to the nearest hundredth. Now we can use the formula

$$V = \frac{1}{3} \cdot B \cdot h = \frac{1}{3} \cdot 132.67 \cdot 8.2 = 362.63.$$

So the volume is 362.63 in³.   ◆

A summary of all the volume formulas discussed in this section is in the table in the margin.

Note that the specific formulas for B = Area of Base are left out for the prism and pyramid because the base for each could be any polygon (rectangle, triangle, etc.).

■ **Surface Area**

Surface area gives the amount of two-dimensional space taken up by the surface of a three-dimensional object. So, for example, if someone wanted to find the surface area of the Earth, they would actually be looking for the flat area of the shape shown in the margin. This is a rather unusual shape to find the area of. Fortunately, the sphere and the cone have formulas for finding surface area, and other objects do not require a formula as we will now explore. Let's look at how we could take a right prism, right cylinder, or right pyramid and find their surface area. *Note:* We will not look at oblique solids in this discussion on surface area.

In order to find the surface area of a triangular prism, you will have to find the area of two triangles and three rectangles. For a cylinder, you must find the area of two circles and one rectangle. Finding the surface area of a rectangular pyramid is tricky; you have to be able to find measurements on its triangular sides. Be careful to distinguish between the height of the pyramid itself and the height(s) of its triangular sides (these are called **lateral heights**). Lateral height is necessary for the surface area of a cone as well. Here are the formulas for the surface area of a sphere and a right cone.

Sphere: $S = 4\pi r^2$

Right cone: $S = \pi \cdot r \cdot \ell + \pi \cdot r^2$

In the formula for the surface area of a right cone, the variable ℓ represents the lateral height of the cone, as opposed to the perpendicular height:

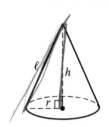

$$h = \text{perpendicular height}$$
$$r = \text{radius}$$
$$\ell = \text{lateral height}$$

Finding the lateral height may require the use of the Pythagorean Theorem.

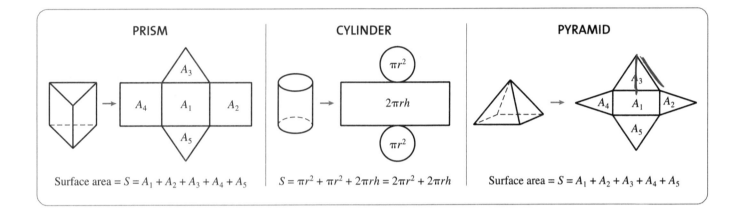

PRISM	CYLINDER	PYRAMID
Surface area $= S = A_1 + A_2 + A_3 + A_4 + A_5$	$S = \pi r^2 + \pi r^2 + 2\pi rh = 2\pi r^2 + 2\pi rh$	Surface area $= S = A_1 + A_2 + A_3 + A_4 + A_5$

EXAMPLE 9-7   The Surface Area of a Cone

Find the surface area of the cone below.

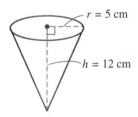

$r = 5$ cm

$h = 12$ cm

◆ **SOLUTION**

Notice that the lateral height is not already given. Fortunately, it is the hypotenuse of a right triangle. This is always the case for lateral height. So, as mentioned above, we need only use the Pythagorean Theorem to find it.

$$\ell^2 = 5^2 + 12^2 \Rightarrow \ell^2 = 169 \Rightarrow \ell = 13$$

Now we can finish finding the surface area easily using the formula given.

$$S = \pi \cdot r \cdot \ell + \pi \cdot r^2$$
$$= \pi \cdot 5 \cdot 13 + \pi(5)^2$$
$$= 65\pi + 25\pi = 90\pi \approx 90(3.14) = 282.6$$

So the cone as a geometric solid has a surface area of 282.6 square centimeters.   ◆

EXAMPLE 9-8   Packaging Food

A company that makes ice cream cones needs to arrange shipping the cones to stores. They have selected a packaging shown below. How many square inches of cardboard will be used to make each package?

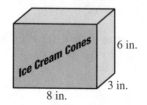

◆ SOLUTION

The box has six faces. Each one is a rectangle. Opposite sides of the box are congruent rectangles, so we should have three rectangular areas each counted twice:

$$S = 2 \cdot \ell \cdot w + 2 \cdot w \cdot h + 2 \cdot \ell \cdot h$$
$$= 2(3)(8) + 2(8)(6) + 2(3)(6)$$
$$= 48 + 96 + 36 = 180$$

Each box will require 180 square inches of cardboard.   ◆

So we see that in some cases finding surface area is easiest by using a formula, and in some cases it can be approached by finding the area of each of the object's sides. This next example illustrates how we can approach trying to find the surface area of a prism.

EXAMPLE 9-9   The Surface Area of an Irregular Prism

Find the surface area of the following prism, given that its base has an area of 50 in².

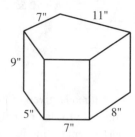

◆ SOLUTION

We know that the base is 50 in², so we just have to find the areas of the five rectangles that make up its lateral sides. Note that the length of each of these rectangles is 9 in., and we are given their widths. So the area of all these rectangles combined is given by the sum $5 \cdot 9 + 7 \cdot 9 + 8 \cdot 9 + 11 \cdot 9 + 7 \cdot 9 = 45 + 63 + 72 + 99 + 63 = 342$. The area of the lateral sides is 342 in². Now we can add the area of the base in twice to get the total surface area: $342 + 50 + 50 = 432$. The total surface area of the prism is 432 in².   ◆

Notice in this last example that when we found the area of the five rectangles involved, this could have been done by finding the perimeter of the base and multiplying by the height. That is, $(5 + 7 + 8 + 11 + 7) \cdot 9 = 38 \cdot 9 = 342$. This works because the lateral sides of the prism could be thought of as one big rectangle.

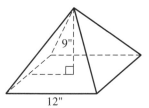

This idea leads to a formula that can be used to help find the surface area of any prism. If we take the perimeter of the base and multiply by the perpendicular height, we get the area of the lateral sides. Then we just add the area of the base two times. The formula looks like this:

Surface Area of a Prism $= S = 2 \cdot B + P \cdot h$, where
B = area of the base, P = perimeter of the base, and h = height of the prism

Finally, we show an example from which no useful formula can efficiently be derived; the point being that even if we don't have a formula we can often work out the surface area using our knowledge of two-dimensional geometry.

EXAMPLE 9–10   Using the Pythagorean Theorem to Help Find the Surface Area of a Pyramid

The following pyramid has a square base. Find the total surface area of the pyramid. Round to the nearest tenth if necessary.

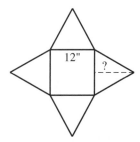

◆ **SOLUTION**

We are trying to find a flat area that looks like this:

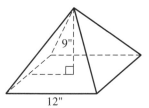

The one problem we have is that we don't know the height of the triangle. The number 9" refers to the height of the pyramid, not the height of its triangular, lateral side.

SUMMARY OF SURFACE AREA FORMULAS	
OBJECT	FORMULA FOR SURFACE AREA
Box	$S = 2\ell w + 2\ell h + 2hw$
Cylinder	$S = 2\pi r^2 + 2\pi rh$
Cone	$S = \pi \cdot r \cdot \ell + \pi \cdot r^2$, where ℓ = lateral height
Sphere	$S = 4\pi r^2$
Prism	$S = 2B + Ph$, where B = area of base and P = perimeter of base

Fortunately, however, the height of the lateral side is the hypotenuse of a right triangle whose legs are the height of the pyramid and half the length of its base. So we find this missing measurement in the calculation below, rounding to the nearest tenth as instructed.

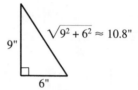

So 10.8" is the height of the lateral side of the pyramid. The surface area then becomes the sum of the square base and four times the area of the triangular lateral face:

$$S = s^2 + 4 \cdot \left(\frac{1}{2} \cdot b \cdot h\right) = 12^2 + 4 \cdot \left(\frac{1}{2} \cdot 12 \cdot 10.8\right) = 144 + 4 \cdot 64.8 = 403.2$$

The surface area of the pyramid is 403.2 in^2.   ◆

Section 9 | Exercises

For exercises 1–4, use the terms from the **Vocabulary Checklist** *to fill in the blanks.*

> ✓ **VOCABULARY CHECKLIST:**
>
> linear unit   volume
>
> square unit   surface area
>
> cubic unit   sphere
>
> solid

1. The _surfac are_ of a three-dimensional object is measured using a square unit.

2. If the definition of a circle is applied in all directions instead of just points in a plane, the result is a

   _____.

3. The volume of a three-dimensional object is measured using a(n) _cubic unit_.

4. The perpendicular height of a three-dimensional object is measured using a(n) _____.

For exercises 5–8, answer TRUE or FALSE.

Ⓣ | F   5. The amount of liquid that can be held by a soda can would be measured using cubic units.

Ⓣ | F   6. In geometry, any three-dimensional object is referred to as a solid.

Ⓣ | F   7. The following statement makes sense: "The total capacity of my new refrigerator is 95 cubic feet."

T | Ⓕ   8. The following statement makes sense: "The surface area of the moon is 38.7 million kilometers."

In exercises 9–19, sketch the indicated figure and, if appropriate, use variables to label the measurements that would be required to find its volume.

9. A right cylinder.

10. A right cone.

11. An oblique cylinder.

12. An oblique pyramid.

13. A sphere.

14. An oblique prism.

15. An oblique cone.

16. An oblique pentagonal prism.

17. A right hexagonal prism.

18. A right solid that cannot be labeled by any name.

19. An oblique solid that cannot be labeled by any name.

20. Find the volume and surface area of the cube.

4 in.

21. Find the volume and surface area of the cube.

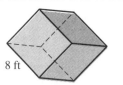

8 ft

384

22. Find the volume and surface area of the rectangular solid.

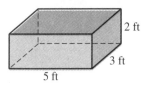

2 ft

3 ft

5 ft

23. Find the volume and surface area of the rectangular solid.

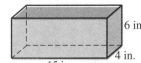

6 in.

4 in.

15 in.

$2(15)(4) + 2(15)(6) + 2(4)(6)$ $348 \, in^2$

24. Find the volume and surface area of the cylinder. Leave π in your answer.

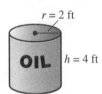

r = 2 ft

OIL   h = 4 ft

25. Find the volume and surface area of the cylinder. Use π ≈ 3.14.

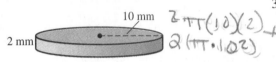

10 mm

2 mm

$2\pi(10)(2) +$
$2(\pi \cdot 10^2)$

753.98

26. Find the volume of the prism.

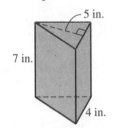

5 in.

7 in.

4 in.

27. Find the volume and surface area of the right prism.

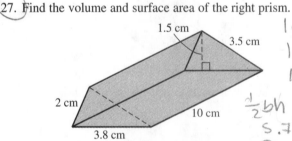

1.5 cm

3.5 cm

2 cm

10 cm

3.8 cm

10×3.8
10×3.5
$10.2 \div 20$
$\frac{1}{2}bh = \frac{1}{2}(3.8)(1.5)$
5.7
$98.7 cm^2$

28. Find the volume.

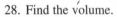

B = 35 m²

h = 7 m

29. Find the volume.

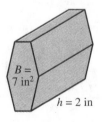

B = 7 in²

h = 2 in

30. Find the volume of the following pyramid, given that its base is rectangular.

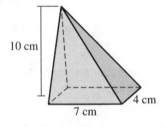

10 cm

4 cm

7 cm

31. Find the volume of the following pyramid, given that its base is rectangular.

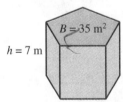

7 in.

8 in.

3 in.

32. Find the volume and surface area of the cone. Use π ≈ 3.14 and round to the nearest hundredth if necessary.

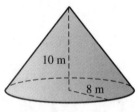

10 m

8 m

33. Find the volume and surface area of the cone. Use π ≈ 3.14 and round to the nearest hundredth if necessary.

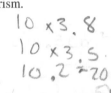

5 m

12 m

$\pi 5^2 + \pi 5 \cdot 13$
282.74

34. Find the volume and surface area. Leave π in your answer.

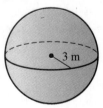

3 m

35. Find the volume and surface area. Use $\pi \approx 3.14$.

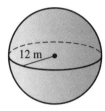

12 m

36. Find the volume. Use $\pi \approx 3.14$.

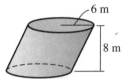

6 m

8 m

37. Find the volume. Leave π in your answer.

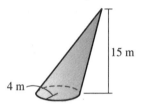

15 m

4 m

38. Find the volume and surface area of the following prism, given that the base has an area of 14 in².

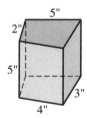

5"

2"

5"

3"

4"

39. The following shape is an oblique trapezoidal prism. Find its volume.

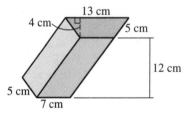

13 cm

4 cm

5 cm

12 cm

5 cm

7 cm

40. Find the volume of the pyramid.

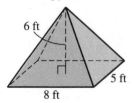

6 ft

5 ft

8 ft

41. Find the surface area of the pyramid. You will need to use the Pythagorean Theorem to find its lateral height.

$64 + \frac{1}{2}(32)(17.04)$

256.66

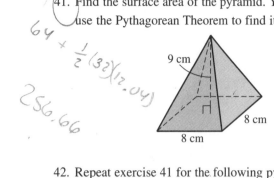

9 cm

8 cm

8 cm

42. Repeat exercise 41 for the following pyramid.

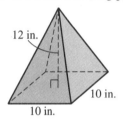

12 in.

10 in.

10 in.

43. Find the volume.

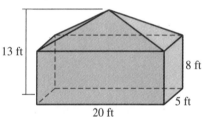

13 ft

8 ft

5 ft

20 ft

44. Find the volume. Round to the nearest hundredth if necessary. (*Hint:* Note the right angle.)

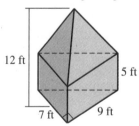

12 ft

5 ft

7 ft

9 ft

45. Find the volume and surface area. Use $\pi \approx 3.14$.

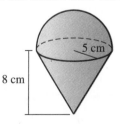

5 cm

8 cm

46. Find the volume. Use $\pi \approx 3.14$.

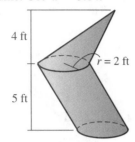

47. The following picture is of a hexagonal pyramid. What is the area of the base if the total volume is 18 cm³?

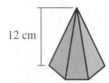

48. What is the height of the following oblique cylinder, given that its total volume is 252π yd³?

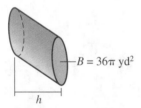

49. **Laying a Foundation.** Wayne needs to lay a concrete foundation with the following dimensions. If one bag of cement makes 2.5 cubic feet of concrete, how many bags will Wayne need to complete the job?

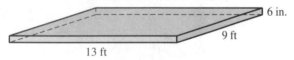

50. A box is twice as long as it is wide, three times as high as it is long. Its width is 4 inches. What is the volume and surface area of the box?

51. **How Much Water Covers the Earth?** The Earth has a diameter of approximately 8000 miles. Given that 70% of the Earth is covered in water, how many square miles of the Earth is covered in water?

52. **Packaging Pens.** A company that makes pens has sheets of cardboard that are 38 square inches. The company wants to use all 38 square inches of cardboard for each box to ship its pens in. Because of the pens, the width must be

$\frac{1}{2}$ inch and the height must be 5 inches. What should the length of each box be so that all the cardboard is used? *Hint:* Start with the formula for the surface area of a box.

53. **Measurements of a Basketball.** A standard basketball has a radius of about 4.5 inches. What is its volume and surface area?

54. **Piling Sand.** A processor cleans sand that will be made into glass and piles it in a cone. If the pile has a diameter of 10 feet and a height of 6 feet, what is its volume? Use $\pi \approx 3.14$.

55. Suppose, as in the previous exercise, that a processor is dumping sand into a pile. Given that the pile stays at a constant height of 21 feet, what must the radius be for the volume to be 1028 cubic feet? Use $\pi \approx \frac{22}{7}$.

56. **Storing Oil.** A cylindrical oil drum has a radius of 1 ft and a height of 3 ft. If 1 cubic foot contains about 7.5 gallons, about how many gallons are in one oil drum? Use $\pi \approx 3.14$.

57. **Medicine Capsules.** A pharmaceutical company is designing medicine capsules. The capsules are cylinders with half spheres on each end. If the length of the cylinder is 12 mm and the radius is 2 mm, how many cubic mm of medication can one capsule hold? Round your answer to the nearest tenth.

58. **Pouring Concrete.** In the rectangular patio area below, concrete needs to be filled in the entire area, excluding the flower bed and the pool. If the concrete is to be poured 6 inches deep, how many cubic feet of concrete will be needed? Use $\pi \approx 3.14$.

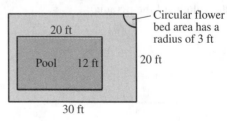

Geometry Project 9 │ Approximating Volumes

For this project you will choose three items and approximate their volume and/or surface area by making certain assumptions and using formulas for familiar shapes. At least one of the items should be an item for which the volume is known (like a drink bottle) so you can see how accurate your approximation is.

For example, you may choose a 2.5 gallon water container like the one shown. Its volume could be approximated using just a rectangular box. Or, for accuracy, you could use several shapes, including boxes, half-cylinders, and triangular prisms, adding and subtracting out volumes as needed. The surface area could be closely approximated using several two-dimensional shapes like rectangles and trapezoids. Our assumption here is that these shapes can be used even though the edges of the container are smooth and rounded. Be creative and resourceful. See if you can find a challenge. Here are some conversions that will help you in seeing how accurate you are.

UNIT	CUBIC INCHES	CUBIC CENTIMETERS
1 gallon	231.0	3785.4
1 liter	61.025	1000.0
1 quart	57.75	946.35
1 pint	28.875	473.18

For each of the three objects, make a report formatted as follows.

Object: _____

Assumptions: _____

Sketches and measurements:

Approximate surface area: _____

Approximate volume: _____

Exact volume (if known): _____

SECTION

10

Trigonometry

Trigonometry (trig) is an entire branch of mathematics founded on the relationships between the sides of right triangles. Trigonometry is one of the most practically useful branches of mathematics and is heavily involved in science (particularly physics). The applications we present here are similar in nature to the type of applications presented in Section 4: More About Triangles—Similarity and Congruence. In fact, you may want to go back and review that section. Some of the work in this section requires the use of a scientific calculator.

Recall the examples in Section 4, where a surveyor was trying to find the height of a tree and we needed to find the length of wire being used to support a tower. The problems were solved by exploiting the relationships between sides in 45-45-90 and 30-60-90 triangles. The 45-45-90 and 30-60-90 triangles are two specific cases in which finding the ratios between sides is relatively easy, using the Pythagorean Theorem.

■ Terminology Associated with Ratios in a Right Triangle

■ Finding Unknown Values

■ Trigonometric Ratio Values for Special Angles

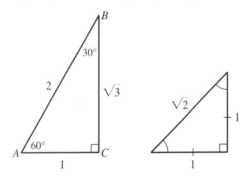

Trigonometry is based on the fact that you can find these kinds of ratios for *any* right triangle if you know the measure of its angles. **If we know the angle measures in a right triangle, we also know how long the sides are in relation to each other.** This can be stated in a way that is a bit more mathematically precise as follows: **The angles in a right triangle determine the ratios between the sides.**

■ Terminology Associated with Ratios in a Right Triangle

Since we are talking about the ratios in a right triangle, we name them.

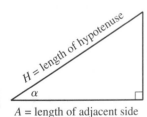

- **Sine** of an angle refers to the ratio of the leg opposite the angle to the hypotenuse of the triangle.
- **Cosine** of an angle refers to the ratio of the leg adjacent to the angle to the hypotenuse of the triangle.
- **Tangent** of an angle refers to the ratio of the leg opposite the angle to the leg adjacent to the angle.

O = length of opposite side

A = length of adjacent side

SECTION

10

Trigonometry

Trigonometry (trig) is an entire branch of mathematics founded on the relationships between the sides of right triangles. Trigonometry is one of the most practically useful branches of mathematics and is heavily involved in science (particularly physics). The applications we present here are similar in nature to the type of applications presented in Section 4: More About Triangles—Similarity and Congruence. In fact, you may want to go back and review that section. Some of the work in this section requires the use of a scientific calculator.

Recall the examples in Section 4, where a surveyor was trying to find the height of a tree and we needed to find the length of wire being used to support a tower. The problems were solved by exploiting the relationships between sides in 45-45-90 and 30-60-90 triangles. The 45-45-90 and 30-60-90 triangles are two specific cases in which finding the ratios between sides is relatively easy, using the Pythagorean Theorem.

■ Terminology Associated with Ratios in a Right Triangle

■ Finding Unknown Values

■ Trigonometric Ratio Values for Special Angles

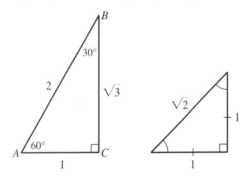

Trigonometry is based on the fact that you can find these kinds of ratios for *any* right triangle if you know the measure of its angles. **If we know the angle measures in a right triangle, we also know how long the sides are in relation to each other.** This can be stated in a way that is a bit more mathematically precise as follows: **The angles in a right triangle determine the ratios between the sides.**

■ Terminology Associated with Ratios in a Right Triangle

Since we are talking about the ratios in a right triangle, we name them.

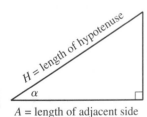

O = length of opposite side

A = length of adjacent side

- **Sine** of an angle refers to the ratio of the leg opposite the angle to the hypotenuse of the triangle.
- **Cosine** of an angle refers to the ratio of the leg adjacent to the angle to the hypotenuse of the triangle.
- **Tangent** of an angle refers to the ratio of the leg opposite the angle to the leg adjacent to the angle.

141

The box in the margin uses variables to illustrate and summarize these definitions. It also gives the notation used for each ratio. As an example, the equation $\sin(30°) = \frac{1}{2}$ is saying that in a 30-60-90 triangle, the ratio of the side opposite the 30° angle to the hypotenuse is always $\frac{1}{2}$. Similarly, $\cos(30°) = \frac{\sqrt{3}}{2}$.

EXAMPLE 10-1   Finding Trigonometric Ratios

The following right triangle has had all its sides and angles labeled. Use the triangle to find these trigonometric ratios: sin 50°, sin 40°, cos 40°, tan 50°.

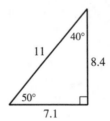

◆ **SOLUTION**

Let's look at each value separately.

- sin 50° refers to the ratio of the leg opposite the 50° angle to the hypotenuse. So
  $\sin 50° = \frac{8.4}{11} \approx 0.76$.
- Notice that for the 40° angle, the opposite leg is the leg measuring 7.1. So
  $\sin 40° = \frac{7.1}{11} \approx 0.65$.
- cos 40° refers to the ratio of the leg adjacent to the 40° angle to the hypotenuse, so
  $\cos 40° = \frac{8.4}{11} \approx 0.76$.
- Tangent is the ratio of the leg opposite the indicated angle to the leg adjacent, so
  $\tan 50° = \frac{8.4}{7.1} \approx 1.18$.   ◆

One important thing to take from this last example is that which leg is "opposite" and which is "adjacent" depends on the angle that has been indicated with the trigonometric ratio. This next example is very similar to the last one, except that we will be forming the trig ratios from a triangle that is mostly labeled with variables.

EXAMPLE 10-2   Completing Trigonometric Equations

Consider the following triangle. Note that the Greek letters alpha (α) and beta (β) have been used as variables to represent two angle measures. Use the triangle to decide what should be in place of the question mark to make each of these trigonometric equations true: $\sin \alpha = \frac{x}{?}$ $\sin \beta = \frac{?}{7}$ $\cos \beta = \frac{x}{?}$ $\tan(?) = \frac{x}{y}$.

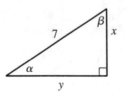

In trigonometry, it is traditional to use Greek letters to signify angles. This helps distinguish them from side lengths, which are usually assigned a letter from our alphabet. We will use the Greek letter alpha (α) as our generic angle measure.

α	ν
β	ξ
γ	ο
δ	π
ε	ρ
ζ	σ
η	τ
θ	υ
ι	φ
κ	χ
λ	ψ
μ	ω

◆ SOLUTION

Let's look at each equation separately.

- $\sin \alpha = \frac{x}{?}$: Sine is the ratio of the leg opposite that angle (which is x) to the hypotenuse (which is 7). So the question mark should be 7 for the equation to be true.

- $\sin \beta = \frac{?}{7}$: The leg opposite the angle β is the leg marked by y. So the question mark should be y.

- $\cos \beta = \frac{x}{?}$: Cosine is the ratio of the leg adjacent to the angle (which is x) to the hypotenuse (which is 7). So the question mark should be 7.

- $\tan(?) = \frac{x}{y}$: Tangent is the ratio of the opposite leg to the adjacent leg. For which angle is x opposite and y adjacent? The answer is angle α. So the question mark should be α.   ◆

This last example is pretty abstract. But the skill it develops is very practical. As mentioned earlier, trigonometry is based on the fact that you can find ratios between sides of a right triangle given one of the non-right angles. So, for example, if you pick up a calculator and punch in tan 54, the calculator will display the number 1.376. You would interpret this as follows:

In a right triangle, if an angle measures 54°, then the ratio of the opposite leg to the adjacent leg must be 1.376.

To someone trying to find the height of an enormous building, this is the most useful thing imaginable. All the person has to do to find the height of the building in the margin is walk away from the building until the angle from their position to the top is 54°, then measure the distance between them and the building. If they do this, they have created the triangle to the right. So they know one of the values making up the tangent ratio. Now it's simple:

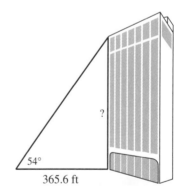

$$\tan(54°) = \frac{O}{A} = \frac{O}{365.6} = 1.376$$
$$\frac{O}{365.6} = 1.376$$
$$O = 503.1$$

So, just like that, the building is 503.1 feet tall.

■ Finding Unknown Values

This process is how we are able to know the distance between the Earth and distant stars, find the height of Mount Everest, or determine the distance to an approaching ship on the horizon. If you want to find any distance, set up a right triangle that has a missing distance as a side. Then all you need is any one angle and any one side and you can find your missing distance easily. In a few moments, we will show you the most powerful applications we have to offer in this book. But first, let's look at an example that is completely abstract to get a better feel for using trig ratios to find missing values on a right triangle. This next example is the classic abstract trig problem: Find the missing measurements in a right triangle, given an angle and a side length.

EXAMPLE 10-3   Solving a Right Triangle

Solve the right triangle. That is, find all the missing values. Round the results to two decimal places.

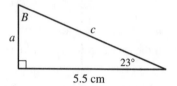

◆ **SOLUTION**

We have been asked to find three missing values: one angle and two sides. Finding the angle is easy. Since the figure is a triangle, it must be true that $23° + 90° + B = 180°$, so $B = 67°$. For the sides, we need to pick a trig ratio that involves the side we already know. Since we were given the angle 23°, let's use the cosine of 23°.

$$\cos(23°) = \frac{\text{Adjacent}}{\text{Hypotenuse}} = \frac{5.5}{c} = 0.92$$
$$\frac{5.5}{c} = 0.92$$
$$5.5 = 0.92c$$
$$c = 5.98$$

To find the other missing side, notice that we could use the Pythagorean Theorem. In practice, feel free to do just that. For the sake of illustrating another trig ratio, we are going to use the tangent in this example.

$$\tan(23°) = \frac{\text{Opposite}}{\text{Adjacent}} = \frac{a}{5.5} = 0.424$$
$$\frac{a}{5.5} = 0.42$$
$$a = 2.31$$

So we have found $B = 67°$, $a = 2.31$ cm, and $c = 5.98$ cm.

To help check our work, we can use the Pythagorean Theorem on the values of a, b, and c to make sure they actually form a right triangle. Since we have been rounding, we are only looking for these to be approximately equal.

$$(2.31)^2 + (5.5)^2 \approx 5.98^2$$
$$5.34 + 30.25 \approx 35.76$$
$$35.59 \approx 35.76$$

This is not a guarantee that our work is correct, but if the equation had not been approximately equal, we would have known we were wrong. ✓ ◆

Once you understand how to find missing values on a triangle, you are ready to answer some pretty impressive questions. Questions like: How far is it from the planet Mars to the sun? This next example illustrates.

EXAMPLE 10–4   Astronomy

In the 1600s, Edmond Halley (as in Halley's comet) found that the distance from the Earth to the sun was about 93,000,000 miles. Suppose that the Earth, the sun, and Mars are aligned so that they form the right triangle shown here. Use the information to find the distance between the planet Mars and the sun, and to find the distance at this particular time from the planet Earth to the planet Mars.

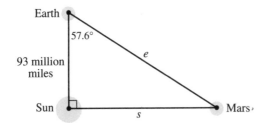

◆ **SOLUTION**

First, notice that the leg of the triangle that is 93 million miles is adjacent to the 57.6° angle. In the figure, we have labeled the distance from the sun to Mars with the variable s. This leg is opposite the 57.6° angle. So, to find s we have the adjacent side and want to find the opposite. Which trig ratio do we use? Tangent! The equation looks like this:

$$\tan 57.6° = \frac{s}{93}$$

$$1.58 = \frac{s}{93}$$

$$s = 146.9$$

So Mars is 146.9 million, or 146,900,000 miles, from the sun.

In the figure, we have labeled the distance from Earth to Mars with the variable e. This is the hypotenuse of the triangle. So, to find e we have the adjacent side and want to find the hypotenuse. Which trig ratio do we use? Cosine! The equation looks like this:

$$\cos 57.6° = \frac{93}{e}$$

$$0.54 = \frac{93}{e}$$

$$0.54e = 93$$

$$e = 172.2$$

So Mars is 172.2 million, or 172,200,000, miles from Earth.   ◆

Our next example shows another reason why the abstract exercise of finding missing measurements of a triangle has very practical and compelling applications. The problem involves an **angle of depression.** An angle of depression measures the angle between a line of sight and the horizon. Some problems also involve an **angle of elevation.**

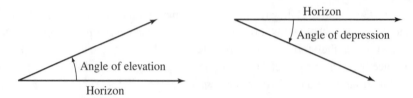

EXAMPLE 10–5   Forest Ranging

Towers are built in some forest areas to watch for fires. A fire tower spots a fire in the distance. A ranger measures that the angle of depression from the top of the tower to the fire is 1.3°. If the tower is 32 meters tall, how far is it from the tower to the fire? Round to the nearest tenth.

◆ **SOLUTION**

The angles in the figure below are not drawn to scale. In truth, 1.3° is a *very* small angle. But the figure will work for solving the problem.

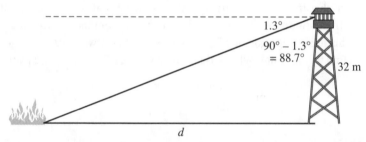

Notice that the angle we were given is not part of the triangle we are using to find the distance d. But because the tower is perpendicular to the horizon, our angle is complementary to the 1.3° angle, which makes it 88.7°. In relation to this angle, we know the adjacent leg and are being asked to find the opposite leg, so we use the tangent. The trig equation becomes

$$\tan 88.7° = \frac{d}{32}$$

$$44.1 = \frac{d}{32}$$

$$d = 1411.2$$

So the fire is 1411.2 meters away from the tower.   ◆

■ Trigonometric Ratio Values for Special Angles

Finally, the following table gives trig ratios for common angles. All these values come straight from the information about 45-45-90 and 30-60-90 triangles. Using these values makes it possible to leave answers to some trig problems in an exact form instead of having to round to a decimal place.

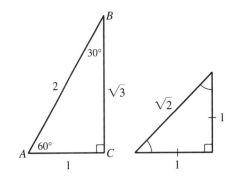

° α	sin α	cos α	tan α
30°	$\frac{1}{2}$	$\frac{\sqrt{3}}{2}$	$\frac{\sqrt{3}}{3}$
45°	$\frac{\sqrt{2}}{2}$	$\frac{\sqrt{2}}{2}$	1
60°	$\frac{\sqrt{3}}{2}$	$\frac{1}{2}$	$\sqrt{3}$

TRIG RATIOS FOR SPECIAL ANGLES

EXAMPLE 10-6   Using Exact Trigonometric Values

Solve the following triangle, giving exact values.

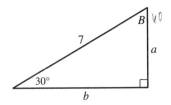

◆ SOLUTION

Since $B + 30° + 90° = 180°$, the measure of angle B is 60°.

Notice, then, that the triangle is a 30-60-90, and we could find the missing side lengths from the techniques we developed for 30-60-90 triangles. Instead, we will use the trigonometric ratios from the table above.

To find side a, for example, we can use

$$\sin 30° = \frac{a}{7}$$

From the table above, we see that

$$\sin 30° = \frac{1}{2}, \text{ so}$$

$$\frac{1}{2} = \frac{a}{7}$$

$$a = 3\frac{1}{2}$$

To find side b we can use

$$\cos 30° = \frac{b}{7}$$

From the table above, we see that

$$\cos 30° = \frac{\sqrt{3}}{2}, \text{ so}$$

$$\frac{\sqrt{3}}{2} = \frac{b}{7}$$

$$b = \frac{7\sqrt{3}}{2}$$ ◆

Section 10 | Exercises

*For exercises 1–4, use the terms from the **Vocabulary Checklist** to fill in the blanks.*

> **VOCABULARY CHECKLIST:**
>
> trigonometry   cosine
>
> ratio   tangent
>
> sine

1. A trigonometric _____ compares two sides of a right triangle.

2. The two trigonometric ratios that involve the hypotenuse are _____ and _____.

3. _____ is a branch of math founded on the study of right triangles.

4. _____ of an angle gives the ratio of the legs in a right triangle that has that angle.

For exercises 5–8, answer TRUE or FALSE.

T | F   5. The sine and cosine ratios are reciprocals of each other.

T | F   6. The trigonometric ratios are only relevant for a *right* triangle.

T | F   7. In a 30-60-90 triangle, the ratio of the side opposite the 30° angle to the hypotenuse must be $\frac{\sqrt{3}}{2}$.

T | F   8. In a 45-45-90 triangle, the ratio of the legs must be 1.

In exercises 9–11, an angle α is given as part of a right triangle. In each problem, find the following: sin(α), cos(α), tan(α).

9.

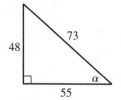

10.

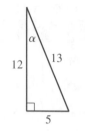

11.

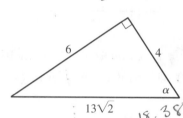

In exercises 12–20, use a calculator to find the value to three decimal places.

12. sin 37°

13. sin 85°

14. cos 56°

15. cos 89°

16. tan 42°

17. tan 67°

18. cos 76.5°

19. tan 19.7°

20. sin 53.9°

Exercises 21–26 refer to the following triangle. Note that the Greek letter beta (β) has been used as a variable to represent an angle. Decide what should be in place of the ? to make the trigonometric equation complete.

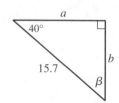

21. $\tan 40° = \dfrac{b}{?}$

22. $\sin 40° = \dfrac{?}{15.7}$

23. $\cos \beta = \dfrac{b}{?}$

24. $\tan \beta = \dfrac{?}{b}$

25. $\cos(?) = \dfrac{a}{15.7}$

26. $\sin(?) = \dfrac{a}{15.7}$

Exercises 27–32 refer to the following triangle. Note that the Greek letters alpha (α) and theta (θ) have been used as variables to represent angles. Decide what should be in place of the ? to make the trigonometric equation complete.

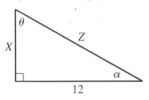

27. $\sin(\alpha) = \dfrac{?}{Z}$

28. $\tan(\alpha) = \dfrac{X}{?}$

29. $\cos(\theta) = \dfrac{X}{?}$

30. $\tan(\theta) = \dfrac{12}{?}$

31. $\cos(?) = \dfrac{12}{Z}$

32. $\sin(?) = \dfrac{12}{Z}$

33. Solve the right triangle. That is, find all the missing values.

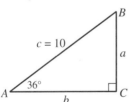

34. Solve the right triangle.

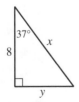

Exercises 35–38 refer to the following right triangle. In each case, given the values, find all other missing values.

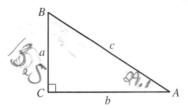

35. $A = 30°$, $c = 10$

36. $B = 72°$, $a = 5$

37. $B = 18.2°$, $b = 5.7$

38. $A = 29.1°$, $a = 13.5$

39. **Finding the Height of a Tower.** The angle of depression of a television tower to a point on the ground 36.0 meters from the bottom of the tower is 29.5°. Find the height of the tower.

40. **Nautical Distance.** The lighthouse in Cape Hatteras, North Carolina, is 191 feet tall. Suppose someone at the top of the lighthouse spots a ship in the distance and the angle of depression to the ship is 16°. How far offshore is the ship?

41. **Nautical Distance.** The world's tallest lighthouse is the Marine Tower in Yokohama, Japan. It stands an impressive 348 feet tall. From the top of the light, suppose the angle of depression to an approaching ship is 9°. How far offshore is the ship?

42. **Surveying.** A person stands 100 feet away from a building and finds the angle of elevation to the top of the building is 78°. To the nearest foot, how tall is the building?

43. **Surveying.** A person stands 78 feet away from a building and finds the angle of elevation to the top of the building is 50°. To the nearest foot, how tall is the building?

44. **Meteorology.** In the golden age of meteorology, the height of clouds was found as follows:

    Shine a strong spotlight straight up at the cloud and mark where it meets the cloud (this will be visible). Move 200 yards away from the spotlight and record the angle of elevation from your position to the point where the light hits the cloud.

If this angle comes out to be 68°, how high is the cloud?

45. **Astronomy.** Use the following figure to find the distance between the planet Jupiter and the sun, and the distance between the Earth and Jupiter.

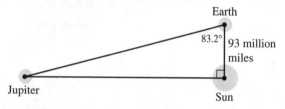

46. Find the missing sides of the figure pictured below. Give exact values. *Hint:* Use the trig ratios for special angles.

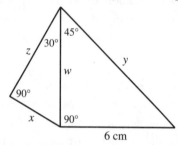

47. Suppose that an equilateral triangle with sides that are 10 cm long (pictured below) is cut exactly in half in such a way that a right angle is formed. Find the area of the shaded triangle.

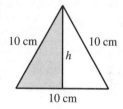

48. Find the height of an isosceles triangle having a base of 184.2 cm if the angle opposite the base is 68°.

49. Fill in the following table.

α	sin α	cos α	tan α
30°			
45°			
60°			

Three more trigonometric ratios can be defined using the reciprocals of the three discussed in this section:

- Cosecant of an angle α is denoted csc(α) and is the reciprocal of the angle's sine.
- Secant of an angle α is denoted sec(α) and is the reciprocal of the angle's cosine.
- Cotangent of an angle α is denoted cot(α) and is the reciprocal of the angle's sine.

In exercises 50–55, find the trig ratios, giving exact values when possible.

50. csc(30°)

51. cot(45°)

52. sec(60°)

53. sec(23°)

54. csc(79°)

55. cot(3°)

| **Geometry Project 10** | Trig Meets World |

The following are real-world problems that require the use of trigonometry and techniques from algebra. We have provided hints to help get you started.

1. Mount Rushmore has a total height of 500 feet. A surveyor has been asked to verify the height of Washington's face. To do this, she obviously can't just make the face part of a single right triangle since none of the measurements would be known.

   Instead, she stands at a far distance and measures the angle of elevation from the ground to the bottom of Washington's chin, finding it to be 19.4°. From the same point, she measures the angle of elevation from the ground to the top of Washington's head, finding it is 21.8°. Use variables and given measurements to label the figure below to illustrate the scenario. Do not bother labeling the hypotenuse.

   Use the figure to create two trigonometric equations. There should be two unknown values, one of which is the height of Washington's head. Use these equations to solve the problem.

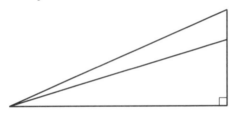

2. The famous Mazatlan lighthouse is situated at the top of this hill (look closely). A surveyor has been asked to verify the height above sea level of the lighthouse.

   To do this, the surveyor stands at a far distance from the lighthouse and measures that the angle of elevation to the top of the lighthouse is 18°. Next the surveyor moves 400 feet closer to the lighthouse and remeasures the angle of elevation, finding that it has increased to 24°. Use variables and given measurements to label the figure below to illustrate the scenario. Do not bother labeling the hypotenuse.

   Use the figure to create two trigonometric equations. There should be two unknown values, one of which is the height above sea level of the lighthouse. Use these equations to solve the problem.

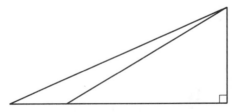

acute angle	An angle whose measure is between 0° and 90°.
acute triangle	A triangle whose interior angles are all acute.
adjacent angles	Two angles that share a side and do not overlap.
algebra	The branch of mathematics in which symbols are used to represent numbers and relationships are explored that hold for all numbers.
alternate angles	Two angles that lie on opposite sides of a transversal.
angle	The space between two rays, lines, or line segments that intersect.
angle bisector	A line, ray, or line segment that divides an angle into two angles of equal measure.
arc	The curve between two points on a circle.
area	The amount of space (measured in square units) taken up by a two-dimensional object.
between point	A point that is collinear to two other points such that the sum of its distance from the two other points is equal to the distance between those points.
bisect	To divide into two equal parts.
center	The point from which all points on a given circle are equidistant.
central angle	The angle formed by two radii on a circle.
chord	A line segment whose endpoints are on the same circle.
circle	The set of all points in a plane that lie a given distance (called the radius) from a given point (called the center).
circumference	The distance (measured in linear units) around a circle.
collinear	Three or more points that lie on the same straight line.
complementary angles	Two angles whose measures add up to 90°.
concave polygon	A polygon for which at least one diagonal falls on the exterior.
congruent	Having the same shape and measure.
convex polygon	A polygon for which all diagonals lie on the interior.
coplanar	Four or more points that lie in the same plane. Also, two or more lines that lie in the same plane.
corresponding angles	Two angles that lie on the same side of a transversal and in the same position relative to the lines being cut by the transversal.
cosine	For a given angle in a right triangle, the ratio of the length of the side adjacent to the angle to the length of the hypotenuse of the triangle.
cubic unit	A measure used for volume.
decagon	A ten-sided polygon.
degree	$\frac{1}{360}$ of one complete revolution.
diagonal	The line segment that connects two nonadjacent vertices in a quadrilateral or polygon.
diameter	A line segment whose endpoints are points on a circle and that passes through the center of the circle. Also, the length of this line segment.
equilateral triangle	A triangle in which all three sides are congruent.
exterior angle	Two angles that lie on the exterior of two lines being cut by a transversal. Also, an angle formed by extending a side of any polygon that lies on the exterior of the polygon.
geometry	The study of the properties, measures, and relationships of points, lines, angles, surfaces, and solids.
heptagon	A seven-sided polygon.
hexagon	A six-sided polygon.
hypotenuse	In a right triangle, the side opposite the right angle.
interior angle	Two angles that lie on the interior of two lines being cut by a transversal. Also, an angle in a polygon whose vertex is a vertex of the polygon.
isosceles trapezoid	A trapezoid for which the legs are congruent.
isosceles triangle	A triangle for which two sides are congruent.

kite	A closed four-sided object with two pairs of adjacent congruent sides.
lateral height	The distance from the top of a cone to a point on its circular base. Also, the height of the triangular side of a pyramid.
leg	In a right triangle, a side that makes up the right angle.
line	In geometry, an undefined object whose existence and nature are understood intuitively.
line segment	The set of all points on a line that lie between two given points called endpoints.
linear unit	A unit of measure used for distance.
median	In a triangle, a line segment drawn from one vertex of a triangle to the opposite side that bisects the side.
midpoint	Given two points, the point that is collinear to them and is equidistant from each one.
n-gon	An *n*-sided polygon.
nonagon	A nine-sided polygon.
noncollinear	Three or more points that do not lie on the same straight line.
oblique solid	A solid whose side(s) make non-right angles with its base.
obtuse angle	An angle whose measure is between 90° and 180°.
obtuse triangle	A triangle that has an obtuse interior angle.
octagon	An eight-sided polygon.
parallel lines	Two lines that exist in the same plane but do not intersect.
parallelogram	A closed four-sided object for which the opposite sides are parallel.
pentagon	A five-sided polygon.
perimeter	The distance (measured in linear units) around an object.
perpendicular	Two rays, line segments, or lines that intersect to form one or more right angles.
pi (π)	The ratio of the circumference to the diameter in any circle; denoted by the Greek letter π.
plane	In geometry, an undefined object whose existence and nature are understood intuitively.
point	In geometry, an undefined object whose existence and nature are understood intuitively.

polygon	A closed figure made of line segments that do not cross.
Pythagorean Theorem	A theorem that states that the sum of the squares of the legs on a right triangle is equal to the square of the hypotenuse.
Pythagorean triple	A set of three natural numbers that satisfy the Pythagorean Theorem.
quadrilateral	A four-sided polygon.
radius (plural: *radii*)	The distance from the center of a circle or sphere to any point on the circle or sphere. Also, a line segment whose endpoints are the center of a circle (or sphere) and a point on the same circle (or sphere).
ratio	A comparison of two quantities written in the same form as a fraction.
ray	All the points on a line that start at one point and extend infinitely in one direction.
rectangle	A closed four-sided object for which all interior angles are right angles.
regular polygon	A polygon for which all sides are congruent.
rhombus	A closed four-sided object in which all sides are congruent.
right angle	An angle whose measure is exactly 90°.
right solid	A solid whose side(s) make right angles with its base.
right triangle	A triangle that has one interior right angle.
same-side angles	Two angles that lie on the same side of a transversal.
scalene triangle	A triangle whose sides all have different lengths.
scaling factor	A number that expresses the relationship between the sides of similar objects.
secant	A line, line segment, or ray that intersects a circle at two points.
sector	The area enclosed by two radii on a circle.
similar	Two or more objects that have the same shape.
sine	For a given angle in a right triangle, the ratio of the length of the side opposite the angle to the length of the hypotenuse of the triangle.
skew lines	Two lines that do not exist in the same plane.
solid	Any three-dimensional object.
sphere	The set of all points in space that lie a given distance (called the radius) from a given point (called the center).

square	A closed four-sided object for which all interior angles are right angles and all sides are congruent.
square unit	A unit of measure used for area.
straight angle	An angle whose measure is exactly 180°.
supplementary angles	Two angles whose measures add up to 180°.
surface area	The amount of space (measured in square units) taken up by the surface of a three-dimensional object.
tangent	For a given angle in a right triangle, the ratio of the length of the side opposite the angle to the length of the side adjacent to the angle.
tangent line	A line that intersects a circle at one point.
tangent segment or ray	A line segment or ray that intersects a circle at one point but, if extended into a line, would not intersect the circle again.

transversal	Any line that intersects two other lines.
trapezoid	A closed four-sided object for which one pair of sides is parallel.
triangle	A three-sided polygon.
trigonometry	A branch of math founded on the study of the relationships between the sides of right triangles.
vertex (plural: vertices)	The endpoints of the line segments that make up a polygon.
vertical angles	Angles that lie on opposite sides of intersecting lines, rays, or line segments.
volume	The amount of space (measured in cubic units) taken up by a three-dimensional object.

SECTION 1 Basic Geometry Concepts

1. ray

3. midpoint, bisector

5. line segment

7. T

9. F

11.

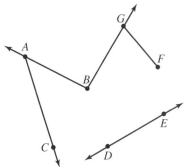

13. $\overleftrightarrow{AS}$, $\overrightarrow{GK}$, $\overleftrightarrow{XV}$, $\angle WFK$, B

15.

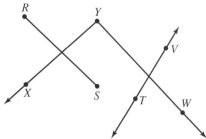

17. $\overline{TP}$, $\overline{PR}$, $\overline{YR}$, $\overline{TY}$, $\overline{TR}$, $\overline{PY}$, $\overline{PM}$, $\overline{YM}$

19. yes.

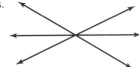

21. no

23. one. Since X is the endpoint, there is only one point three inches away in the direction of the ray.

25. Answers will vary.

27. Answers will vary.

29. 24 angles

31. 6. Drawings will vary.

33. 6. Drawings will vary.

35. F and H

37. E and F or E and C

39. 3

41. No. It violates the given definition of *between*.

43. 13.7

45. $GH = 3$

47. $DH = 9$

49. $GH = 4$

51. We should have $DG + GH = DH$, but $12 + 3 \neq 18$.

53. $LE = 20$, $ED = 20$, $LD = 40$

55. $\overline{AD} \cong \overline{JP}$; $\overline{AL} \cong \overline{LP} \cong \overline{DJ}$

57. 3.25 in; 2.9 in; 6.15 in

59. 28.95 in

61. 16 in, 9 in, 18 in

63. B

65. yes

67. a. 6.5; b. 0; c. 3.625

SECTION 2 More about Angles

1. vertical, adjacent, supplementary

3. transversal

5. perpendicular

7. F

9. F

11. Mars—right, Venus—obtuse, Saturn—obtuse, Jupiter—acute

13. 180°

15. 135°

17. 120°

19. 90°

21. $\angle ABC$—acute, $\angle BCA$—acute, $\angle CAB$—right

23. a.

b.

c.

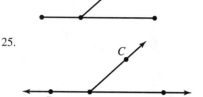

25.

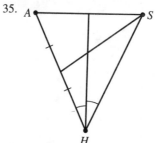

27. a. 130°—obtuse; b. 100°—obtuse; c. 125°—obtuse
   d. 85°—acute; e. 35°—acute

29. $\angle a$ and $\angle d$, $\angle a$ and $\angle b$, $\angle b$ and $\angle c$, $\angle c$ and $\angle d$

31. $m\angle a = m\angle c = 60°$, $m\angle b = m\angle d = 120°$,

33. a. $x = 24$; b. $x = 32$; c. $x = 44$

35. two congruent, 74°

37. two congruent, 53.4°

39. a. neither; b. complementary; c. supplementary; d. neither

41. 141°

43. 9.6°

45. $111\frac{2}{3}°$

47. 157.5°

49. 112°

51. If two numbers are both less than 90, they cannot add up to 180.

53. $\angle BCE$, $\angle ABE$, $\angle BAD$, $\angle ACD$

55. Answers will vary.

57. $\angle ADC$ and $\angle CDE$

59. Answers will vary.

61. $\overline{AD}$ and $\overline{DE}$

63. $x = 26$

65.

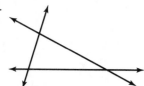

67. All obtuse angles are 107° and all acute angles are 73°.

69. All obtuse angles are 110° and all acute angles are 70°.

71. H, F, N, E, Z

73. F, E

75. A, H, F, N, E, Z

SECTION 3 Triangles

1. isosceles

3. right triangle

5. hypotenuse

7. T

9. T

11. $\triangle XYZ$

13. $\overline{XY}$, $\overline{YZ}$, $\overline{ZX}$

15. isosceles

17. $\angle Z$

19. scalene triangle

21. obtuse triangle

23. right triangle

25. isosceles right triangle

27. b and c are impossible

29. obtuse triangle

31. isosceles triangle

33. $\triangle DJR$—acute isosceles triangle, $\triangle SJR$—obtuse triangle, $\triangle DRS$—right triangle

35.

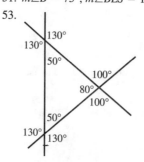

37. 10°

39. 45°

41. $60\frac{8}{15}$

43. 35°, 87°, 58°

45. 36°, 36°, 108°

47. $x = 19$

49. $x = 38\frac{4}{5}°$

51. $m\angle B = 75°$, $m\angle BLS = 150°$

53.

55. $c = 10$ cm

57. $c = \sqrt{18}$ in ≈ 4.2 in

59. $a = \sqrt{32}$ in ≈ 5.6 in

61. $\sqrt{9650}$ ft ≈ 98.2 ft

63. $\sqrt{180.5}$ in ≈ 13.4 in

65. 1. $\overline{AL}$ should be the longest side.

   2. $\overline{AB}$ and $\overline{BL}$ should be congruent.

   3. The triangle is draw as right but it does not satisfy the Pythagorean theorem.

67. The sum of any two sides must be greater than the third side, but $2 + 3$ is less than 6.

SECTION 4 Similarity and Congruence

1. scaling factor

3. similar triangles

5. F

7. F

9. The first implies the triangles are congruent: congruent angles and congruent sides. The second implies the triangles are similar: congruent angles and proportional sides.

11. $\triangle BAC \cong \triangle EDF$; $\overline{AB} \cong \overline{ED}$; $\overline{FD} \cong \overline{AC}$; $\angle B \cong \angle E$; $\angle D \cong \angle A$

13. $\triangle NDY \sim \triangle NLK$; $\angle LKN \cong \angle DYN$; $\angle NDY \cong \angle NLK$;

$\dfrac{LK}{YD} = \dfrac{LN}{ND} = \dfrac{KN}{YN}$

15. $\triangle HSB \cong \triangle ALB$; $\angle H \cong \angle A$, $\angle S \cong \angle L$, $\angle HBS \cong \angle ABL$; $\overline{HB} \cong \overline{AB}$, $\overline{HS} \cong \overline{AL}$, $\overline{SB} \cong \overline{LB}$

17. $\triangle BEM \sim \triangle BWJ$; $\angle J \cong \angle M$, $\angle W \cong \angle E$, $\angle JBW \cong \angle MEB$;

$\dfrac{JW}{EM} = \dfrac{JB}{BM} = \dfrac{BW}{BE}$

19. ASA theorem for congruence

21. AA theorem for similarity. $\triangle ABC \sim \triangle EDC$

23. AA theorem for similarity. $\triangle ABD \approx \triangle ACE$

25. $f = 20$, $b = 7$

27. $x = 2\frac{2}{3}$, $y = 5\frac{1}{3}$

29. 11.5 ft

31. Nancy is taller since she has a longer shadow. Nancy would be 5'8". Michelle would be 5'3".

33. 5 ft

35. The scaling factor is 3. The area of the larger triangle is 9 times the area of the smaller.

37. $7, 7\sqrt{2}$

39. $10, 10\sqrt{3}$

41.

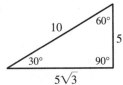

43.

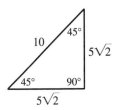

45. $54\sqrt{3} \approx 93.5$ ft

SECTION 5 Quadrilaterals

1. diagonal

3. kite

5. F

7. F

9. Drawings will vary.

11. Drawings will vary.

13. Drawings will vary.

15. Drawings will vary.

17. square, rectangle

19. kite

21. rectangle

23. sometimes

25. sometimes

27. sometimes

29. rectangle, square, parallelogram, rhombus, kite

31. kite, rectangle, square, parallelogram, rhombus

33. kite

35.

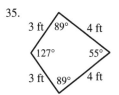

37.

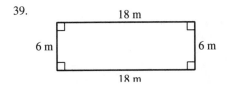

39.

41.

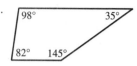

43. $x = 15$

45. 75° and 105°

47.

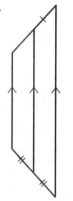

49. 40°

51. 80°

53. 5 cm

55.

SECTION 6 Polygons

1. hexagon

3. 24-gon

5. T

7. T

9. a. yes; b. no; c. no; d. yes; e. no; f. yes

11. Drawings will vary.

13. Drawings will vary.

15. Drawings will vary.

17. Drawings will vary.

19. Drawings will vary.

21. 3, 4, or 6

23. hexagon

25. isosceles triangle, rectangle, decagon, square

27. 144°

29. 1080°

31. 6 diagonals

33. 90°

35. 25 sides

37. 7740°

39. 64 diagonals

41. 2.5°

43. 9 diagonals

45. 20 diagonals

47. $n = \dfrac{360}{180 - A}$. Given the measure of one angle in a regular polygon, this formula will give the number of sides.

49. 10 sides

51. $x = 95°, y = 75°$

53. $x = 133°$

SECTION 7 Area and Perimeter

1. square unit

3. area

5. F

7. F

9. about 13 cm²

11. about 20 km²

13. $P = 18$ ft, $A = 20.25$ ft²

15. $P = 56$ ft, $A = 75$ ft²

17. $P = 20\frac{1}{3}$ m, $A = 23\frac{1}{3}$ m²

19. $P = 13.9$ in, $A = 6.46$ in²

21. $P = 20$ cm, $A = 16$ cm²

23. $P = 19$ cm, $A = 16.08$ cm²

25. $P = 17\frac{8}{15}$ m, $A = 14\frac{14}{45}$ m²

27. $P = 42$ ft, $A = 96$ ft²

29. $A = 144$ cm²

31. $P = 10$ in, $A = 4.2$ in

33. $A = 66$ in²

35. 36 in²

37. $P \approx 950$ mi, $A \approx 43{,}750$ mi²

39. $P \approx 1100$ mi, $A \approx 67{,}500$ mi²

41. $\ell = 9$ m, $w = 7$ m

43. 6 in, 18 in, 18 in

45. Drawings will vary.

47. Drawings will vary.

49. Drawings will vary.

51. $9800

53. $121

55. $285, 19 bags

57. $238

59. $3960

61. $35,496

SECTION 8 Circles

1. arc
3. radius
5. tangent
7. F
9. F
11. a. 12 ft; b. $6\frac{4}{5}$ ft; c. 11.6 mi; d. $(2b + 10)$ cm; e. $(14x)$ m
13. radius: $\overline{CB}$, $\overline{CD}$, $\overline{CA}$; diameter: $\overline{AD}$
15. There are 8 radii and 3 diameters.
17. Drawings will vary.
19. $m\angle TBK = 110°$, $m\angle HBT = 180°$, $m\angle HBK = 70°$
21. $C = 10\pi$ in, $A = 25\pi$ in^2
23. $C = 44$ in, $A = 154$ in^2
25. $C = 7.6\pi$ in, $A = 14.44\pi$ in^2
27. $r = 6$ m
29. $r = 15$ mi
31. $r = 5$ yd
33. $r = 2$
35. $A = 26.28$ ft^2. 20.28 ft of trim would be needed.
37. $A = 83.04$ ft^2
39. $A = 15.25$ in^2
41. one 13-inch pizza
43. Jasmine: 45°, Symphony: 22.5°
45. 1.83 ft
47. 90°
49. 120 cm
51. 72°
53. 18 cm
55. $x = 2$
57. $x = 2$
59. $P = 10.71$ ft, $A = 7.07$ ft^2
61. The width is 2 yd. The area is 138.16 yd^2
63. $P = 51.98$ in., $A = 184.93$ in^2
65. 113 hours
67. The circumference is doubled. The is increased by a factor of four.

SECTION 9 Volume and Surface Area

1. surface area
3. cubic unit
5. T
7. T
9. Drawings will vary.
11. Drawings will vary.
13. Drawings will vary.
15. Drawings will vary.
17. Drawings will vary.
19. Drawings will vary.
21. $SA = 384$ ft^2, $V = 512$ ft^3
23. $SA = 348$ in^2, $V = 360$ in^3
25. $SA = 325.6$ mm^2, $V = 628$ mm^3
27. $SA = 98.7$ cm^2, $V = 28.5$ cm^3
29. $V = 14$ in^3
31. $V = 56$ in^3
33. $SA = 282.2$ m^2, $V = 314$ m^3
35. $SA = 1808.64$ m^2, $V = 7234.56$ m^3
37. $V = 80\pi$ m^3
39. $V = 480$ cm^3
41. $SA = 221.44$ cm^2
43. $V = 966\frac{2}{3}$ ft^3
45. $SA = 305.05$ cm^2, $V = 471$ cm^3
47. $B = 4.5$ cm^2
49. 24 bags
51. 140,672,000 mi^2
53. $SA = 254.34$ in^2, $V = 381.51$ in^3
55. 6.84 ft
57. 184.2 mm^3

SECTION 10 Trigonometry

1. ratio
3. trigonometry
5. F
7. F
9. $\sin \alpha = \frac{48}{73}$, $\cos \alpha = \frac{55}{73}$, $\tan \alpha = \frac{48}{55}$
11. $\sin \alpha = \frac{3\sqrt{2}}{13}$, $\cos \alpha = \frac{2\sqrt{2}}{13}$, $\tan \alpha = \frac{3}{2}$
13. 0.996
15. 0.017
17. 2.356
19. 0.358
21. a
23. 15.7
25. 40°
27. X
29. Z
31. α
33. $a = 5.88$, $b = 8.09$
35. $B = 60°$, $a = 5$, $b = 5\sqrt{3}$
37. $A = 71.8°$, $a = 17.34$, $c = 18.25$
39. 20.37 m

41. 2,197.19 ft

43. 93 ft

45. 779.92 million miles, 785.45 million miles

47. $A = 25\sqrt{3}$ cm^2

49.

α	$\sin \alpha$	$\cos \alpha$	$\tan \alpha$
30°	$\frac{1}{2}$	$\frac{\sqrt{3}}{2}$	$\frac{\sqrt{3}}{3}$
45°	$\frac{\sqrt{2}}{2}$	$\frac{\sqrt{2}}{2}$	1
60°	$\frac{\sqrt{3}}{2}$	$\frac{1}{2}$	$\sqrt{3}$

51. 1

53. 1.09

55. 19.08